Intellectual Property in Consumer Electronics,
Software and Technology Startups

Gerald B. Halt, Jr. · John C. Donch, Jr.
Robert Fesnak · Amber R. Stiles

Intellectual Property in Consumer Electronics, Software and Technology Startups

 Springer

Gerald B. Halt, Jr.
John C. Donch, Jr.
Amber R. Stiles
Volpe and Koenig, P.C.
Philadelphia
USA

Robert Fesnak
Fesnak and Associates, LLP
Blue Bell, PA
USA

ISBN 978-1-4939-4796-6   ISBN 978-1-4614-7912-3   (eBook)
DOI 10.1007/978-1-4614-7912-3
Springer New York Heidelberg Dordrecht London

Printed on acid-free paper

Springer is part of Springer Science+Business Media (www.springer.com)

Acknowledgments

The authors would like to thank everyone at the law firm of Volpe and Koenig, P.C. for providing support and encouragement in preparing this book. The authors would like to thank Fesnak and associates for their assistance on the valuation methodologies chapter. The authors would also like to individually thank Anthony S. Volpe for his contribution to the intellectual property enforcement and litigation chapter, and C. Frederick Koenig III for his contribution to the copyright chapter.

Contents

Introduction

Creativity can create economic value. This maxim holds as true for the electrical technology industry—which comprises, for the purposes of this book, a broad spectrum of companies that deal in consumer electronics, computers, wireless technology, electrical components, etc.—as it does for other industries. Such value may come from a new innovation, edging out competitors in a market, creating a revenue stream where there was none, or increasing market reputation. This book provides an introduction to intellectual property law as applied to the electrical technology industry. This area of law provides the legal framework for bridging creativity and the value that may come from it. Through the proper use of intellectual property law, one has a much better chance of transforming creativity into economic value.

Intellectual property law recognizes a creator's rights in ideas, innovations, and goodwill. Being intangible, intellectual property differs from real property (land) or personal property (physical possessions) that are secured, controlled, and protected using physical means such as fences, locks, alarms, and guards. Because intellectual property is a product of the mind, there is often no easy way to build a "fence" around it. Consider some of the most valuable trademarks in the world: Apple® and Microsoft®. These companies could not protect these marks with a physical fence. It is intellectual property law that provides a legal fence of trademark protection to protect the goodwill of the two trademarks. Consider also the numerous novel products—software and hardware—that these technology giants regularly introduce to the market after investing significantly in research and development. Intellectual property, in the form of patents, plays a key role maximizing the companies' revenue from these investments, by protecting them from the market impact of knock-off products, by encouraging mutually beneficial cooperation with competitors through licensing deals, and in some cases, by providing a company with bargaining chips that can be used to avoid litigation.

There are many intellectual property pitfalls that await the unwary. Different rules apply to different types of intellectual property (IP) and you may forfeit your rights if you do not take the appropriate measures to secure and protect them. It is important to understand the types of IP protection and the respective rules that govern each type of IP.

1. *Patent*: Patents may be granted for the invention of any new and useful process, machine, manufacture or composition of matter, or any new useful improvement thereof. A patent is a property right that grants the inventor or owner the right to exclude others from making, using, selling, or offering to sell the invention as defined by the patent's claims in the United States for a limited period of time.
2. *Trademark*: A trademark is a word, phrase, symbol, or design, or combination of words, phrases, symbols, or designs which identifies and distinguishes the source of the goods or services of one party from those of others. Trademarks promote competition by giving products corporate identity and marketing leverage.
3. *Copyright*: Copyrights protect original works of authorship fixed in a tangible medium of expression. Copyrighted works include literary, dramatic, and musical compositions, movies, pictures, paintings, sculptures, computer programs, etc. Copyright protects the expression of an idea, but not the idea itself.
4. *Trade Secret*: Generally, a trade secret is any formula, manufacturing process, method of business, technical know-how, etc., that gives its holder a competitive advantage and is not generally known. The legal definition of a trade secret and the protection afforded to a trade secret owner varies from state to state.

Table xvii highlights some of the attributes of and distinctions between these different types of IPs:

Table xvii Different types of IPs-attributes and distinctions

	Patent	Trade Secret	Trademark	Copyright
Underlying theory	Limited monopoly to encourage innovation in exchange for disclosure of invention to the public	Protects proprietary and sensitive business information against improper acquisition	Used to identify the source of a product or service to consumers, and to distinguish the source of products or services from other sources	Limited monopoly to encourage the authorship of works
Subject matter	Processes, machines, articles of manufacture, compositions of matter, asexually reproduced plants, designs for articles of manufacture. Laws of nature, mathematical algorithms, natural phenomena, mental steps, etc. are not patentable	Formulas, patterns, compilations, programs, devices, methods, techniques, processes, etc., that derive independent economic value from being "secret"	Trademarks, service marks, trade names, certification marks, collective marks, trade dress	Literary, musical, choreographic, dramatic, and artistic works *limited by* idea/expression dichotomy (no protection for ideas, systems, methods, procedures); no protection for facts/research
Legal source	Patent Act (35 U.S.C. § 100 *et seq.*)	State statutes (e.g., Uniform Trade Secrets Act); common law	Lanham Act (15 U.S.C. §§ 1051–1127); common law	Copyright Act (17 U.S.C. 101 *et seq.*); some limited common law
Legal standards	Must be patentable subject matter, novel, non-obvious, and useful	Information must not be generally known or readily available. Reasonable efforts to maintain secrecy must be taken	Must be distinctive or carry a secondary meaning (for descriptive and geographic marks), and must be used in commerce	Must be an original work of authorship fixed in a tangible medium
Scope of rights	Exclusive right to prevent others from making, using, selling or offering to sell the subject matter of the patent	Protection against improper acquisition by others	Exclusive right to use the mark within a particular territory depending on the type of trademark protection	Exclusive right to perform, display, reproduce, or make derivative works

(continued)

Table xvii (continued)

	Patent	Trade Secret	Trademark	Copyright
Term	20 years from application filing date	No time limitation. Protection is available as long as kept secret	No time limitation. Protection is available as long as used in commerce	Generally, the term is the life of the author plus 70 years. For works of corporate authorship, the term is 120 years after creation or 95 years after publication, whichever endpoint is earlier
Enforcement/ Remedies	File suit for patent infringement. Remedy can be damages (lost profits or reasonable royalty) and injunctive relief	File suit for misappropriation, conversion, or breach of contract. Remedy is typically damages	File suit for trademark infringement. Remedies can include injunctive relief, accounting for profits, destruction of goods, etc.	File suit for infringement. Remedies include injunctive relief, destruction of infringing goods, and damages (actual or profits or statutory damages)

Patents, trademarks, trade secrets, and copyrights all have a strong presence in the electrical industry. Patents and trademarks are the most common means of IP protection in this industry.

As evidenced by the steady stream of novel high-tech offerings and the rate at which the latest gadgets become "obsolete," companies in the electrical industry invest millions in research and development to meet the immense demand for new, useful technologies, and to help bring to market the next wave of cutting-edge products for consumers. The fruits of this research and development—which gives rise to the technologies themselves or to a key component of a particular products' functionality—are typically protected through patents. The following Table xx shows a sampling of some electrical technology classes of patents and how many have been issued by the United States Patent and Trademark Office (USPTO).

Table xx Electrical classes of patents and number of issued patents by USPTO

Class	Class title	Pre-2000	2000	2001	2002	2003	2004	2005	2006	2007	2008	2009	2010	All years
320	Electricity: battery or capacitor charging or discharging	2191	361	397	355	295	251	208	183	186	246	503	623	5799
323	Electricity: power supply or regulation systems	3286	305	344	327	292	277	294	400	384	375	491	576	7341
326	Electronic digital logic circuitry	4562	545	596	596	523	562	517	550	667	658	578	749	11103
336	Inductor devices	1421	127	136	179	203	184	243	222	205	185	152	257	3514
338	Electrical resistors	2139	103	100	105	80	53	50	97	69	47	49	97	2989
345	Computer graphics processing and selective visual display systems	11088	1635	1876	1866	1885	2171	2007	2940	2488	2352	2594	3322	36224
349	Liquid crystal cells, elements, and systems	4454	653	621	628	859	1007	993	951	842	997	1409	1692	15106
717	Data processing: Software development, installation, and management	1406	378	513	341	347	421	481	763	717	714	829	1314	8224

Companies also invest millions in advertising and marketing their brands in order to build up goodwill and consumer loyalty toward their products. Many of the "supermarks"—i.e., trademarks that have achieved a level of popularity to be considered a household name—come from various branches of the electrical technology industry, including AT&T®, Apple®, GE®, Google®, Intel®, Microsoft®, Samsung®, and Sony®. Marks such as these are among the world's most valuable and are instantly recognized by the general consuming public worldwide as a designation of source and associated with an expected level of quality. Indeed, the electrical industry is characterized by significant brand loyalty, and many purchasing decisions are based on brand name alone.

In short, it is very important for a company operating in the electrical industry to secure patent (and, where appropriate, trade secret) rights in its research and development, and to invest in brand and reputation by filing for trademark protection. Once secured, a company can enforce its intellectual property rights against a competitor. Several notable examples are summarized below.

(1) In *NTP, Inc. v. Research-In-Motion*, Case No. 03-1615, 418 F.3d. 1282 (Fed. Cir. 2005), NTP Inc. (NTP), owner of several patents relating to the integration of electronic mail with radio frequency wireless communication networks, sued Research in Motion (RIM), which is the Canadian seller of BlackBerry®, in the United States alleging that elements of the BlackBerry® system infringe various method and system claims of NTP's patents. The Court of Appeals for the Federal Circuit (CAFC) reversed the district court's finding of direct infringement under section 271(a) of an asserted method claim because at least one of the asserted patented method steps occurred in Canada. The CAFC, however, affirmed the district court's ruling of infringement of certain system claims notwithstanding the fact that a key component of the claimed system was located outside the United States.

(2) In *Ebay Inc. v. Mercexchange LLC*, 547 U.S. 388, 126 S.Ct. 1837 (2006), eBay the online auction site, practiced online auction technology for which MercExchange owned patents, including a patent which covers eBay's "Buy it Now" function. Initiating negotiations, eBay sought to outright purchase MercExchange's online auction patent portfolio in 2000, but then abandoned the negotiations. MercExchange sued eBay for patent infringement. The U.S. Supreme Court unanimously decided that an injunction should not automatically issue based on a finding of patent infringement. Furthermore, an injunction should not be denied purely on the basis that the plaintiff does not practice the patented invention. Rather, a federal court must weigh what the Court described as the four factors traditionally used to determine if an injunction should issue.

(3) In *KSR Int'l Co. v. Teleflex*, 550 U.S. 398 (2007), Teleflex, Inc. sued KSR International, claiming that a KSR product infringed Teleflex's patent on connecting an adjustable vehicle control pedal to an electronic throttle control. KSR argued that the claim was not patentable because the combination of the two elements was obvious. The district court ruled in favor of KSR, but the

Court of Appeals for the Federal Circuit reversed. The Supreme Court unanimously reversed the judgment of the Federal Circuit, holding that the disputed claim of the patent was obvious under the requirements of 35 U.S.C. § 103 and that in "rejecting the District Court's rulings, the Court of Appeals analyzed the issue in a narrow, rigid manner inconsistent with § 103 and our precedents," referring to the Federal Circuit's application of the "teaching-suggestion-motivation" (TSM) test.

The above cases were filed in federal court to enforce federal IP rights. Another commonly used option is to file suit in the International Trade Commission (ITC) to prevent the importation of articles that infringe a valid and enforceable U.S. patent, registered copyright, or trademark. For example, in 2009, a U.S. company claimed that Research-in-Motion's BlackBerry devices, servers, and desktop software infringe a patent for an authentication system that can be used with services conducted over the Internet in a complaint to the ITC. *In the Matter of Certain Authentication Systems, Including Software and Handheld Electronic Devices*, Complaint No. 2699. Research-in-Motion is a Canadian company. Ultimately, after five months of negotiations, the parties reached a settlement agreement.

This book illustrates how intellectual property rights can apply by presenting examples from real-world companies. Case studies are used throughout the book to demonstrate how intellectual property rights, management, and business all work together in industry. The examples throughout the book will relate to well-known companies and products that many readers will be familiar with.

Part I of this book provides a comprehensive overview of the most common forms of intellectual property rights. Part II provides guidelines for how high-tech consumer electronics and computer software technology companies can properly secure and implement their intellectual property rights. Part III discusses different monetization strategies, including but not limited to the enforcement, leveraging, and licensing of IP rights.

Examples and Illustrations Throughout this Book

Examples and illustrations will be used throughout this book. Some are real-world scenarios featuring well-known companies. Others are hypothetical scenarios. These are intended to enhance the reader's understanding of the subject matter contained in the chapters.

Part I
Overview of Intellectual Property Rights

Chapter 1
Patents

1.1 Why Apply for a Patent and How Will it Help my Business?

Congress shall have power... To promote the progress of science and useful arts, by securing for limited times to authors and inventors the exclusive right to their respective writings and discoveries.

United States Constitution, Article I, Sect. 8.

A U.S. Patent is a contract between the United States and the inventor(s) in which the owner is granted a limited monopoly to exclude others from making, using, selling, offering for sale, or importing a patented invention into the United States during the term of the patent. In exchange for these exclusive rights, the inventor is required to disclose the full and complete details of the invention to the public. The theory behind the patent system is that if the public has access to complete inventive disclosures, it will develop new and better ways of solving the same problems.

The patent monopoly has some limitations. A patent does not give an owner the right to make, use, or sell an invention. For example, a patent owner can be prevented from selling its patented invention if a competitor's earlier patent covers some part of the patented invention. Further, a U.S. Patent is not enforceable outside the United States; each country offers its own patent protections within its borders.

The patent right to exclude others from making, using, selling, offering for sale, or importing the patented invention creates barriers for competitors to enter the market. Such barriers often facilitate licensing arrangements where some of the patent rights can be separated in licensing arrangements. For example, a company can grant a license to one company to *make* a patented invention, while granting a second license to a second company to *use or sell* the patented invention. Developing a strong portfolio of patent rights (i.e. barriers to entry) can be attractive to investors or may create new business opportunities by reducing the risks of competition.

G. B. Halt, Jr. et al., *Intellectual Property in Consumer Electronics,*
Software and Technology Startups, DOI: 10.1007/978-1-4614-7912-3_1,
© Springer Science+Business Media New York 2014

Example: Apple holds patents on much of the technology incorporated inside of its iPad and iPhone products. However, the centerpiece for each of these devices lies in the core processor. For many years Apple has licensed the technology for the core processor chip from ARM Holdings. ARM is an IP holding company that focuses on semiconductor technology. The chip that is used in iPad 2 and the iPhone 4S is based on ARM's dual-core ARM Cortex-A9 MPCore central processing unit and a dual core PowerVR SGX543MP2 graphics processing unit.

While Apple has the right to exclude others from making, using, selling, offering for sale, or importing iPads and iPhones within the scope of Apple's patent rights, Apple in turn needs a license from ARM to use ARM's processors in Apple products. If Apple were to obtain the core processor from another source for use in new models of the iPad or iPhone, Apple will want assurances from its vender that the new processors are covered by an appropriate license.

1.2 Patentability Requirements

In order for something to be patentable, it must be: (1) patentable subject matter; (2) useful; (3) novel; and (4) non-obvious.

1.2.1 Patentable Subject Matter

Pursuant to the patent statute, "[w]hoever invents or discovers any new and useful process, machine, manufacture, or composition of matter, or any new and useful improvement thereof, may obtain a patent therefore, subject to the conditions and requirements of this title."[1]

A "process" is a way to produce a result. For example, a process may consist of mixing certain ceramic elements at a particular pressure and temperature to create a new ceramic composite. Not all processes are patentable. For example, a pure mathematical algorithm is not patentable. However, a mathematical algorithm included in a process used to determine a useful, concrete and tangible result will in most circumstances be considered patentable subject matter. A "machine" is a device with assembled parts that move to perform a desired operation. A "manufacture" or "article of manufacture" is typically regarded as a man-made, tangible

---

[1]   35 U.S.C. § 101.

object that is not naturally-occurring. A "composition of matter" is any compound, substance, mixture, etc. that is the result of combining two or more ingredients.

Based on the above definitions, it is no surprise that patentable subject matter has been said to "include anything under the sun that is made by man."[2] There are, however, some recognized exceptions including: (1) laws of nature, (2) natural phenomena, and (3) abstract ideas.

Inventions may often encompass more than one category of patentable subject matter. Accordingly, patents will often have more than one type of claim.

Example: Consider the following hypothetical situation. During product testing of a new battery, a senior engineer discovered that the battery would operate at a cooler temperature, thus reducing the risk of the battery reaching interrupt temperature and improving battery life, if an impurity were introduced into the chemical cells of the battery during manufacture process of the battery. The engineer/inventor decides to patent the invention.

In this example, the engineer may be able to pursue protection for both the battery and the method of making the battery. A patent with product claims may give the inventor broader protection because the claims would give the inventor the right to exclude competitors from making the product according to any method claimed in the patent. Method claims are often desirable because even if the product is not held to be novel, the method of making the product may still be novel.

1.2.2 Utility Requirement

A patent application must also demonstrate that the claimed invention is "useful" for some purpose to meet the utility requirement. In most technical fields, this utility requirement has a low threshold easily satisfied by demonstrating any useful result. For a patented invention to fail to satisfy the utility requirement it must be "totally incapable of achieving a useful result," which is rare in applications for processes, machines, and articles of manufacture.

While rare in those instances, failure to satisfy the utility requirement is more common in biotechnology and chemical applications. In the biotechnology and chemical fields, the USPTO typically requires that applications disclose a practical or real-world benefit available from the invention; in other words, a specific,

---

[2] Diamond v. Chakrabarty, 447 U.S. 303 (1980).

substantial and credible utility. Specific utility requires that the applicant have knowledge of what the invention does. Credible utility requires that the claimed invention be believable based on current state of the art. Finally, substantial utility requires that the claimed invention have a real world benefit (e.g., a treatment for a disease). In the chemical field, claims may be rejected for lack of utility if a compound or reaction creates a reasonable doubt as to whether there is a credible utility.

1.2.3 Novelty Requirement

In order for an invention to be patentable, it must be new or "novel" (i.e., not in the prior art). If the prior art shows every element of a claim, the claim is unpatentable as "anticipated" by the prior art. In the U.S., prior art is "everything" in the public domain that existed before the filing date of a patent application. In order for a patented invention to be rejected over a prior art reference, the reference must have been public somewhere in the world. Secret or non-public materials cannot act as prior art. Rules regarding prior art differ around the world. For most foreign countries, prior art is "everything" prior to the priority filing date of a patent application (i.e., most countries do not recognize a "one year" grace period).

The AIA Prior Invention—35 U.S.C. § 102(a)(1) and § 102(a)(2)
Under § 102(a)(1), if an invention was known, used, on sale or was disclosed in a printed publication anywhere in the world before the *effective filing date of invention*, it is not patentable. This section of the AIA expands prior art to include uses and sales anywhere in the world, not just the U.S. However, there are a few exceptions regarding disclosures to the public, which are codified in § 102(b). Section 102(a)(1) also considers prior art to be public use, on sale or "otherwise available to the public before the effective filing date of the claimed invention." For example, displaying a product at a trade show anywhere in the world is likely to be considered a use that bars patentability.

> *Example*: A designer presents a new touch screen smartphone product with fingerprint scanning capability at a technology convention on June 1, 2013, and the designer decides to file a patent application on August 1, 2013. But, another person published an article disclosing the same product on July 1, 2013. During examination of the designer's patent application, a USPTO examiner could rely on the publication as anticipating the designer's patent application claims.

Section 102(a)(2), in essence, switches the U.S. patent application system from a "first-to-invent" to a "first-to-file" regime by indicating that a patent is not available if there is an earlier-filed application describing the claimed invention that "names another inventor."

Statutory Bars—35 U.S.C. § 102 (b)

Under the AIA, § 102(b) offers several exceptions to § 102(a). Any public disclosure, use, offer for sale, or sale of the invention, made by the inventor, joint inventor, or another who obtained the subject matter of the invention directly or indirectly from the inventor is not considered prior art under this section so long as the disclosure is within 1 year of the effective filing date of the patent application. Public disclosure of an incomplete invention may not rise to a statutory bar.

Under the AIA, the inventor's own actions will not result in a § 102(b) statutory bar so long as an application is filed within 1 year of the inventor's act of disclosure. For example, an inventor's public disclosure of the invention at a trade show or offer to sell the invention to anyone is a permissible public disclosure only if an application is filed within a year of that public disclosure. More than 1 year prior to the application filing date can be a statutory bar. An exception to the public use statutory bar is if the invention is being publicly used for bona fide testing or evaluation.

Example: Assume that two inventors, Dr. Fuse and Dr. Diode equally contribute to making a new semiconductor material. Dr. Diode is interested in publishing an article for the electronics industry to disclose their new breakthrough material. If the two inventors intend to pursue patent protection for their invention, Dr. Diode should wait until after the patent application is filed to publish his article. If Dr. Diode publishes his article prior to the date the application is filed, however, the two inventors will have one year from the first date of circulation of the publication to file a U.S. patent application on the semiconductor material. By publishing the article prior to the application's filing date, the inventors may be prohibited from filing foreign patent applications.

Example: During the research and development phase of the smartphone touch screen design, a company wants to test the sensitivity of the touch screen to pressure and touch to determine whether adjustments need to be made. In doing so, the company goes to the local high school and allows students to test the product under the condition that the students agree to complete an evaluation related to the product. After the product testing is complete and the company makes a determination that the product's sensitivity range is acceptable, the company also makes a competitor's product available to the students at the high school for a one-day-only "demo" in order to evaluate their preference between the company's product, and the competitor's product. One year and one day after this preference-testing, the company files a patent application for the product.

In the above example, the company will likely be able to argue that the sensitivity testing is not a statutory bar because it was conducted for bona fide experimental purposes in order to determine whether the touch screen's sensitivity range is acceptable. In contrast, the "preference-testing" evaluation will likely create a statutory bar that prevents patentability of the product because evaluating consumer preference is typically not considered an experimental purpose.

The following chart summarizes the types of materials and acts considered "prior art":

What	Who	Where	When
The invention is publicly known	By another	Anywhere	Before the applicant's effective filing date
The invention is publicly used	By another	Anywhere	Before the applicant's effective filing date
The invention is described in a patent	By another	Anywhere	Before the applicant's effective filing date
The invention is on sale	By anyone	Anywhere	Before the applicant's effective filing date
The invention is described in a publicly available printed publication	By another	Anywhere	Before the applicant's effective filing date
The invention is described in a patent	By anyone	Anywhere	Before the applicant's effective filing date
The invention is otherwise made available to the public	By anyone	Anywhere	Before the applicant's effective filing date
The invention is publicly used	By inventor, joint inventor, or another who obtained the subject matter of the invention directly or indirectly from the inventor	Anywhere	More than 1 year prior to the application effective filing date

(continued)

(continued)

What	Who	Where	When
The invention is on sale	By the inventor, joint inventor, or another who obtained the subject matter of the invention directly or indirectly from the inventor	Anywhere	More than 1 year prior to the application effective filing date
The invention is otherwise made available to the public	By the inventor, joint inventor, or another who obtained the subject matter of the invention directly or indirectly from the inventor	Anywhere	More than 1 year prior to the application effective filing date

1.2.4 Non-Obvious Requirement

An invention is obvious if the differences between the subject matter sought to be patented and the prior art are such that the subject matter as a whole would have been obvious at the time the invention was made to a person having ordinary skill in the art to which said subject matter pertains. For example, merely substituting a screw for a nail would normally not be patentable, since both are commonly used fasteners.

In conducting an obviousness analysis, an examiner may combine multiple prior art references. The examiner cannot, however, combine references arbitrarily. The non-obviousness requirement requires that an examiner step into the shoes of a person of ordinary skill at the time the invention was made and determine whether the claimed invention would have been obvious without using hindsight obtained by reviewing the patent application.

Every obviousness determination considers four factual inquiries: (1) the scope and content of the prior art; (2) the differences between the prior art and the claimed invention; (3) the level of ordinary skill in the pertinent art field at the time of the invention; and (4) objective evidence of obviousness or non-obviousness ("secondary considerations").

The scope and content of the prior art includes art that is directed to the same field of invention as claimed in a patent application, and any other art that is logically relied upon. The prior art used in determining whether an invention is obvious is the same material defined as "prior art" under 35 U.S.C. § 102. Using the above example, if an invention is directed to a touch screen smartphone with fingerprint recognition capability, an examiner might look to the touch screen art, cell phone art, fingerprint recognition software art, and any other art concerned with combining touch screens with cell phones or fingerprint scanning and reading.

Determining the differences between the prior art and the claimed invention is a useful starting point to determine whether the claimed invention would have been obvious in view of the prior art. If the differences between the prior art and claims are trivial, the claimed invention will likely be unpatentable as obvious in view of the prior art.

The level of skill required of a hypothetical person having ordinary skill in the art is more than an ordinary layperson but less than an expert in the field of the invention. Determining the level of skill in the art is a factual question that is often open to debate. Factors that are often considered in such a determination can include the level of sophistication in the technology, the education of ordinary person in the field, and prior art attempts to solve related problems.

Courts refer to objective evidence of obviousness or non-obviousness as "secondary considerations." Such secondary considerations include: long felt need for the invention, commercial success of the invention, and copying by others. For example, if there was a long need for the claimed solution to a problem, or if the invention is commercially successful, the claimed invention is likely not obvious. Also, showing that the prior art teaches away from the claimed invention can be used to support non-obviousness.

1.3 Types of Patents and Patent Applications

There are several types of United States patents issued by the United States Patent and Trademark Office (USPTO): utility, design, and plant patent. Utility patents are the most common and protect functional innovations including "any new and useful process, machine, manufacture, or composition of matter, or any new and useful improvement thereof." Utility patents protect the structure or function of an invention for a term of 20 years from their earliest effective filing date.

Design patents protect "any new, original, and ornamental design for an article of manufacture" for a term of 14 years from their issue date. The subject matter of a design patent may relate to the configuration or shape of an article, to the surface ornamentation on an article, or to both. If a design is primarily the result of an article's function, a utility patent may be preferable over a design patent. For example, the following patents illustrate both a utility patent and a design patent for an Apple iPhone.

U.S. Utility Patent No. 7,869,206

(12) **United States Patent**   (10) **Patent No.:**   **US 7,869,206 B2**
      Dabov et al.   (45) **Date of Patent:**   **Jan. 11, 2011**

(54) **HANDHELD COMPUTING DEVICE**

(75) Inventors: **Teodor Dabov**, San Francisco, CA (US);
                 Hui Leng Lim, San Jose, CA (US); **Kyle**
                 Yeates, Palo Alto, CA (US); **Stephen**
                 Brian Lynch, Portola Valley, CA (US)

(73) Assignee: **Apple Inc.**, Cupertino, CA (US)

(*) Notice: Subject to any disclaimer, the term of this
               patent is extended or adjusted under 35
               U.S.C. 154(b) by 139 days.

(21) Appl. No.: **12/205,826**

(22) Filed: **Sep. 5, 2008**

(65)   **Prior Publication Data**

      US 2010/0061055 A1   Mar. 11, 2010

(51) **Int. Cl.**
      G06F 1/16   (2006.01)
(52) **U.S. Cl.** **361/679.55**
(58) **Field of Classification Search** 361/679.55,
                                                      361/679.56
      See application file for complete search history.

(56)   **References Cited**

              U.S. PATENT DOCUMENTS

5,128,829 A	7/1992	Loew
5,568,358 A	10/1996	Nelson et al.
5,737,183 A	4/1998	Kobayashi et al.
5,796,575 A	8/1998	Podwalny et al.
6,137,890 A	10/2000	Markow
6,153,834 A	11/2000	Cole et al.
6,427,017 B1	7/2002	Toki
6,746,797 B2	6/2004	Benson et al.
6,757,157 B2	6/2004	Lammintaus et al.
6,781,824 B2	8/2004	Krieger et al.
6,847,522 B2 *	1/2005	Fan et al. 361/679.55
6,929,879 B2	8/2005	Yamazaki
7,149,557 B2	12/2006	Chadha
7,236,357 B2 *	6/2007	Chen 361/679.55
7,515,431 B1 *	4/2009	Zadesky et al. 361/752

7,558,054 B1	7/2009	Prest et al.
7,558,057 B1	7/2009	Naksen et al.
7,583,987 B2	9/2009	Park
7,663,607 B2	2/2010	Hotelling et al.
7,688,574 B2 *	3/2010	Zadesky et al. 361/679.21
7,697,281 B2 *	4/2010	Dabov et al. 361/679.55
2002/0102870 A1	8/2002	Burns et al.
2002/0107044 A1	8/2002	Kuwata et al.
2002/0114143 A1	8/2002	Morrison et al.
2003/0081392 A1	5/2003	Cady et al.
2004/0203518 A1	10/2004	Zheng et al.

(Continued)

FOREIGN PATENT DOCUMENTS

EP   1 732 230 A2   12/2006

(Continued)

OTHER PUBLICATIONS

Office Action dated Sep. 30, 2009 in U.S. Appl. No. 12/205,824.

(Continued)

Primary Examiner—Lisa Lea-Edmonds
(74) *Attorney, Agent, or Firm*—Beyer Law Group LLP

(57)   **ABSTRACT**

A minimum Z height handheld electronic device and methods
of assembly is described. The electronic device includes a
single seamless housing having a front opening and a cover
disposed within the front opening and attached to the seamless housing without a bezel.

27 Claims, 28 Drawing Sheets

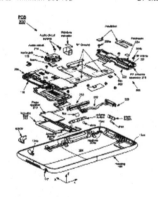

U.S. Design Patent No. D593,087

(12) **United States Design Patent**   (10) **Patent No.:**   **US D593,087 S**
Andre et al.   (45) **Date of Patent:**   ★★ **May 26, 2009**

(54) **ELECTRONIC DEVICE**

(75) Inventors: **Bartley K. Andre**, Menlo Park, CA
(US); **Daniel J. Coster**, San Francisco,
CA (US); **Daniele De Iuliis**, San
Francisco, CA (US); **Richard P.
Howarth**, San Francisco, CA (US);
Jonathan P. Ive, San Francisco, CA
(US); **Steve Jobs**, Palo Alto, CA (US);
Duncan Robert Kerr, San Francisco,
CA (US); **Shin Nishibori**, San
Francisco, CA (US); **Matthew Dean
Rohrbach**, San Francisco, CA (US);
Douglas B. Satzger, Menlo Park, CA
(US); **Calvin Q. Seid**, Palo Alto, CA
(US); **Christopher J. Stringer**, Portola
Valley, CA (US); **Eugene Antony
Whang**, San Francisco, CA (US); **Rico
Zorkendorfer**, San Francisco, CA (US)

(73) Assignee: **Apple Inc.**, Cupertino, CA (US)

(**) Term: **14 Years**

(21) Appl. No.: **29/282,833**

(22) Filed: **Jul. 30, 2007**

Related U.S. Application Data

(63) Continuation of application No. 29/270,880, filed on
Jan. 5, 2007, now Pat. No. Des. 558,756.

(51) LOC (9) Cl. **14-03**
(52) U.S. Cl. .... **D14/341**; D14/203.7; D14/138 G
(58) **Field of Classification Search** D14/137,
D14/138, 147, 191, 218, 247–248, 341–347,
D14/496, 138 R, 138 AA, 138 AB, 138 AC,
D14/138 AD, 138 C, 138 G; D10/65, 78,
D10/104; D13/168; D18/7; 455/556.1,
455/566, 575.1, 575.3; 345/169
See application file for complete search history.

(56)   **References Cited**

U.S. PATENT DOCUMENTS

D289,873 S	5/1987	Gemmell et al.
D337,569 S	7/1993	Kando
D420,354 S *	2/2000	Morales D14/191
D424,535 S	5/2000	Peltola
D456,023 S	4/2002	Andre et al.
D489,731 S	5/2004	Huang
D498,754 S	11/2004	Blyth
D499,423 S	12/2004	Bahroocha et al.
D502,173 S	2/2005	Jung et al.
D504,889 S	5/2005	Andre et al.
D505,950 S	6/2005	Summit et al.
D507,003 S	7/2005	Pai et al.
D514,121 S	1/2006	Johnson
D514,590 S	2/2006	Naruki
D519,116 S	4/2006	Tanaka et al.
D519,523 S	4/2006	Chiu et al.
D520,020 S	5/2006	Senda et al.
D528,542 S	9/2006	Luminosu et al
D528,561 S	9/2006	Ka-Wei et al.
D529,045 S	9/2006	Shin
D532,791 S	11/2006	Kim
D534,143 S	12/2006	Lheem
D535,281 S	1/2007	Yang
D536,691 S	2/2007	Park
D536,962 S *	2/2007	Tanner D9/424
D538,822 S	3/2007	Andre et al.
D541,298 S	4/2007	Andre et al.
D541,299 S	4/2007	Andre et al.
D541,785 S *	5/2007	Hwang et al. D14/138
D546,313 S	7/2007	Lheem
D548,732 S	8/2007	Cebe et al.
D548,747 S	8/2007	Andre et al.
D554,098 S *	10/2007	Lee D14/138
D556,211 S	11/2007	Howard
D557,238 S	12/2007	Kim
7,303,424 B2 *	12/2007	Tu et al. 439/372
D558,460 S *	1/2008	Yu et al. D6/308
D558,756 S	1/2008	Andre et al.
D558,757 S	1/2008	Andre et al
D558,758 S *	1/2008	Andre et al. D14/341
D558,792 S	1/2008	Chigira
D560,683 S *	1/2008	Lee D14/496
D560,686 S	1/2008	Kim et al.
D561,153 S	2/2008	Hong et al.
D561,204 S	2/2008	Toh

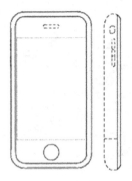

In the above examples, the utility patent (left) protects functional aspects of the iPhone, such as a "liquid crystal display (LCD)" and "an LCD controller." In contrast, the design patent (right) provides a different scope of protection directed to the appearance of the iPhone shown above.

Plant patents are granted to an inventor who "asexually reproduces any distinct and new variety of plant, including cultivated sprouts, mutants, hybrids, and newly found seedlings, other than a tuber propagated plant or a plant found in an uncultivated state" for a term of 20 years after its earliest effective filing date. Asexual reproduction means to reproduce a plant without using seed, and includes techniques such as grafting, budding, or using cuttings, layering, or division in order to assure that offspring are substantially identical to the parent. Naturally occurring plant varieties, however, are not patentable.

1.4 Provisional Patent Applications

In advance of filing a non-provisional or regular patent application, a provisional patent application may be filed to preserve an early filing date for 1 year. The requirements for filing a provisional application include: a specification, drawing figures (if necessary to an understanding of the invention), the official filing fee, and the name and home residence of each inventor. A provisional patent application is not examined by the USPTO. A provisional application may be converted into a non-provisional patent application at any time during this 12 month period. In addition, an applicant has 1 year from its provisional patent application filing date to file any foreign patent applications claiming priority to the provisional patent application filing date.

The benefits of a provisional application are lower costs, the rights to an earlier effective filing date, and minimal filing requirements. Provisional patent applications remain confidential (and a potential trade secret) if the 12 months period lapses and the applicant decides not to pursue a non-provisional patent application.

Example: Returning to the Dr. Diode example, assume that Dr. Diode inadvertently forgot to inform the company he works for that he submitted a description for the new semiconductor material to the *Circuitry Circular Journal* for publication. *Circuitry Circular* will publish tomorrow, and the company wants to file a patent application to preserve its rights to file a foreign patent application before the publication.

Because *Circuitry Circular* publishes tomorrow, it is unlikely that the company and its patent attorney will have sufficient time to prepare a thorough non-provisional patent application. Under this scenario, the company should file a provisional application with as much data as it can possibly submit. The company will then have one year to file a non-provisional patent application, or any foreign patent applications claiming priority to its provisional application's filing date.

1.5 Non-Provisional Patent Applications

Once the decision is made to pursue patent protection for an invention, a patent application should be filed with the United States Patent and Trademark Office (USPTO). The USPTO assigns a filing date and application serial number to the application. The filing date is important because it sets a date which "prior art" references must predate in order to reject the claims of the application.

A regular or non-provisional patent application must provide a specification that: (1) describes the invention in sufficient detail to show one skilled in the art that the inventor possessed the claimed invention at the time of filing ("written description requirement"); and (2) describes the invention in a manner that would allow one skilled in the art to make and use the claimed invention without undue experimentation (e.g. "enablement requirement"). The specification must conclude with one or more claims that particularly point out and distinctly claim the novel subject matter of the invention. Drawings are "necessary for the understanding of the subject matter sought to be patented."[3] Finally, the application must include an oath or declaration naming the true and correct inventors, and must include the requisite filing fee.

1.5.1 Claims

A patent application's claims are critical to defining the scope of protection sought in a patent. The claim's scope has been described as defining the "metes and bounds" of the patented invention. These "metes and bounds" define a patent holder's rights to exclude others from making, using, selling, offering to sell, or importing an accused invention. Thus, if an accused invention falls within a patent claim's scope, it infringes the patent's scope. Before reaching infringement, however, the claims must meet certain requirements.

The claims must be supported by the specification. If the specification describes parts of an invention that are not defined in the claims, it is possible that such disclosure will be dedicated to the public. For this reason, the claims must particularly point out and distinctly claim the novel subject matter of the invention, and should describe the invention as broadly possible based on the specification.

---

[3] 37 C.F.R. § 1.81(a).

The claims in a patent application are typically structured to include independent claims that broadly define the claimed invention, and dependent claims that limit the scope of the independent claims. A dependent claim includes all of the limitation of an independent claim, but includes additional elements that further limit the independent claim. For example, a dependent claim may read, "The apparatus of claim 1, further comprising [additional elements]."

> *Example* of an Independent Claim from U.S. Patent No. 8,026,565
> 1. A thin film semiconductor device including
> a substrate; and
> a number of thin film active layers of inorganic material that are deposited in layers on the substrate and wherein at least one active layer is printed onto one of the substrate and an underlying active layer.
> *Example* of a Dependent Claim from U.S. Patent No. 8,026,565
> 3. A semiconductor device as claimed in claim 1, wherein the substrate is of a material including cellulose.

1.5.2 Specification

The specification of a patent is the written description of the invention. It serves as a disclosure of the invention to the public. If the patent issues and becomes involved in litigation and a term in the claims is ambiguous, the court will look to the specification for guidance on how to construe the claim, (i.e., how the court will interpret the claim). There are several required sections that must be included in the specification. These sections include:

- *The title of the invention* The essence of the invention should be captured in as few words as possible. The title is limited to 500 characters.
- *Cross-reference to related applications* U.S. patent applications filed before March 16, 2013 are required to list any provisional applications which the application is claiming domestic priority from, as well as whether the application is a continuation, continuation-in-part or divisional application from another earlier filed application. In this section, the applicant must identify the provisional application or the parent application by its application number. For U.S. patent applications filed after March 16, 2013, domestic priority claims are set forth in an application data sheet, rather than in the specification.
- *Statement regarding federally sponsored research (if applicable)* If the invention is the result of federally sponsored research funding, the U.S. requires that this be disclosed in the patent application. If the invention is not the result of federally funded research, this section can be excluded from the specification.

- *Abstract* Each application must include an abstract of the invention. The abstract may not exceed 150 words.
- *Background of the invention* This section of the application discusses select, relevant art in the field and emphasizes the major differences between the art and the invention being claimed in the application. The purpose is to point out precisely any and all improvements and challenges which are overcome by the invention.
- *Summary of the invention* The summary is separate and distinct from the abstract of the invention. The summary focuses on the claimed invention. The summary is sometimes used to discuss problems that exist with the prior art and is often used to highlight the advantages that the present invention has over the prior art.
- *Description of the drawings* This section provides a general overview roadmap to navigating the drawings that are included in the application. This section identifies each figure, and provides a brief description of each.
- *Detailed description of the invention* This section is the heart of the patent application. An applicant provides a detailed description of the invention; its characteristics, preferred embodiments, definitions of terms and specific examples of how to practice the invention are provided. The examples illustrate how the invention is meant to operate, and do not limit the claims of the invention in any way.
- *Sequence listing (if applicable)* A sequence listing is not very common because it is only a requirement for those applications that include nucleic acid or amino acid sequences. If sequences are disclosed in the specification, then they need to be included in a listing for easy reference.

Aside from these required sections, the specification must also satisfy three requirements set forth in the first paragraph of 35 U.S.C. § 112: the specification must contain a *written description* of the invention, and of the manner and process of making and using it, in such full, clear, concise, and *exact terms as to enable* any person skilled in the art to which it pertains, or with which it is most nearly connected, to make and use the same.

Written Description Requirement
The specification must fully describe the invention recited in the claims with particularity. While the specification does not have to describe the claims verbatim, it must describe the claimed invention in such a manner that a person of ordinary skill in the application's technical field would understand the claimed invention. Also, the specification should describe as many alternate embodiments of the invention as reasonably permissible in order to avoid any rejections by the USPTO for lack of written description.

Example: A patent application specification that describes a smartphone with fingerprint recognition technology is filed by a company. After filing the patent application, the company learns that consumers prefer a smartphone with voice recognition capability, and would like to pursue patent protection for this embodiment. Because the company's original patent application did not disclose a smartphone with voice recognition capability, any claims to such an embodiment will likely be rejected as not being supported by the specification.

Enablement Requirement

The enablement requirement requires that the specification, at the time the application is filed, describe the invention in such a manner that a person of ordinary skill in the art could *make* and *use* the claimed invention without undue experimentation. The fact that a person of ordinary skill in the art is required to perform *some* experimentation when carrying out the claimed invention does not mean that such experimentation is "undue." However, the quality or quantity of any such experimentation must not be unreasonable or unduly burdensome.

Example: Dr. Fuse has developed a new ceramic material with a unique chemical composition that results in a capacitor dielectric with all the advantages of a traditional ceramic dielectric, but with none of the disadvantages. Dr. Fuse found that baking the ceramic at a low temperature in a pressurized atmosphere gave the new ceramic material enhanced qualities. In a patent application for the new ceramic material, Dr. Fuse openly discloses the specifics of the ceramic formulation in the specification of the patent application; Dr. Fuse also discloses the use of a pressurized atmosphere in the production of the ceramic material, but Dr. Fuse guards the details of the pressurization process as if they were a trade secret. Since Dr. Fuse failed to disclose the details of how to use the pressurized atmosphere in the production process of the ceramic, a person of ordinary skill in the relevant art would not be able to determine the exact temperature and pressure to use without undue experimentation. Dr. Fuse has not provided sufficient description because the disclosure is not sufficient to enable others to practice the invention.

1.5.3 Inventorship

A U.S. patent application must be filed by the actual inventor(s) of the subject matter. Determination of inventorship can be a difficult task that requires legal analysis. "Conception" of the invention is typically considered the key for determining inventorship. Conception is the mental formulation and disclosure by the inventor or inventors of a complete idea for a product or process. Mere contributions of labor or supervision are typically insufficient to vest inventorship rights in the invention. In contrast, in the academic setting, it is often discretionary to name contributors of a research project on published articles. However, naming inventors of a patent application is not discretionary. If the inventorship on an issued patent is incorrect, a court may invalidate the patent.

Example: Two of senior electrical engineers, Dr. Fuse and Dr. Diode, equally contributed to the conception of a new electronic device product. Two entry-level electrochemical engineers, Dr. Lithium and Dr. Ni-Cad, initially tested the product to determine the optimal thermal interrupt temperature of the product's battery under the direction of Dr. Fuse. After Drs. Fuse and Diode accepted the product, Dr. Lithium discovered that the battery would operate at a cooler temperature, thus reducing the risk of the battery reaching interrupt temperature and improving battery life, if an impurity were introduced into the chemical cells of the battery during manufacture of the battery. The company that employs Drs. Fuse, Diode, Lithium and Ni-Cad decides to file a patent application for this invention and needs to determine the inventor(s).

In this example, if the company decides to pursue claims in a patent application directed to the composition of the new electronic device, Dr. Fuse and Dr. Diode should be considered the inventors. If the company decides to pursue claims to a method of making the electronic device that involves the improved battery then Dr. Fuse, Dr. Diode, and design engineer Dr. Lithium should be identified as the inventors.

1.5.4 When Should You Apply for a Patent Application?

Previously, the USPTO granted a U.S. patent to the first inventor to invent. However, patent reform legislation was passed as the Leahy-Smith America Invents Act (AIA) on September 16, 2011. This bill converted the U.S. patent system from a "first-to-invent" into a "first-to-file" system effective on all patent applications filed on or after March 16, 2013. The change aligns the U.S. patent laws more closely with the patent practices of the rest of the world. However, certain critical differences still exist. Because some applications that are filed are still able to claim a priority date before March 16, 2013, these applications still qualify for treatment under the first-to-invent rules. Other applications that are filed after March 16, 2013 that cannot claim an earlier priority data are governed by the first-to-file rules. The two sets of rules are very different in some ways, and will simultaneously exist for the next 20 years.

Patent Applications Filed Before March 16, 2013

Before the AIA transition to a "first-to-file" system, inventors operated under the prior laws of a first-to-invent U.S. patent application system. Under the first-to-invent system, inventors would file a patent application as soon as the invention was complete. An invention was considered to be complete after it is conceived and reduced to practice.

Conception is an inventor's mental formulation and disclosure of a complete idea for a product or process. The test for conception is whether the inventor had an idea that was definite and permanent enough that a person skilled in the art could understand the invention. Completion of an invention's second requirement, reduction to practice, has two types: "constructive" and "actual." Filing a complete patent application satisfies a "constructive" reduction to practice. To prove "actual" reduction to practice, an inventor must have: (1) constructed an embodiment of the invention, and (2) tested the device or process so as to establish its capacity to successfully perform its intended purpose.

After conception, it is important that the inventor act diligently to reduce an invention to practice and file a patent application with the USPTO. The USPTO gives the application its filing date for determining the date that any "prior art references" must precede in order to be cited against the claims in the application. A prior art reference is anything that was publicly available prior to the date of invention. The application filing date is also useful for establishing a priority date over other similar or competing patent applications. For patent applications claiming a priority date under the "old" patent system, the filing date is not the final arbiter of which of two competing applications is entitled to a patent. For example, it may be possible to "swear behind" the application filing date by showing an earlier date of actual reduction to practice, or an early date of conception coupled with diligent reduction to practice. Similarly, during prosecution of a patent application, an applicant can similarly "swear behind" a prior art reference.

Example: Dr. Fuse and Dr. Diode conceived of a new touch screen smartphone product on January 1, 2010. Shortly after Dr. Fuse and Dr. Diode conceived of the new touch screen smartphone, Dr. Fuse resigned from the company, and began working for a competitor. Dr. Diode continued working on the product and completed a working embodiment of the touch screen smartphone on May 1, 2010, and filed a patent application for the product on June 1, 2010. Dr. Fuse learned that the competitor also began marketing a similar touch screen smartphone product shortly after Dr. Fuse was hired. The competitor, through Dr. Fuse, reduced its product to practice on March 1, 2010, and filed a patent application for the product on April 1, 2010. Dr. Diode and his employer want to know whether they can claim priority over the competitor's patent application.

Timeline of Events				
Company's	Competitor's	Competitor's	Company's	Company's
Invention's Conception 01/01/2010	Reduction To Practice 03/01/2010	Application Filing Date 04/01/2010	Reduction To Practice 05/01/2010	Application Filing Date 06/01/2010

Because the competitor reduced its invention to practice prior to the company, the competitor is considered to have priority. However, if the company can provide sufficient evidence, such as an inventor's notebook, that proves it acted diligently to reduce its invention to practice between its January 1, 2010 date of conception and its May 1, 2010 date of reduction to practice, it may be able to claim priority over the competitor's patent application.

Patent Applications Filed After March 16, 2013

Since the U.S. switched to a first-to-file system, the best strategy is to file as soon as possible to secure the earliest effective filing date. The first inventor to file an application is entitled to examination and the chance to obtain the grant of a patent.

1.5.5 Patent Examination

After filing a patent application, the USPTO assigns the application to a patent examiner for examination. During examination, the examiner insures that the application satisfies all formal requirements for the specification, claims, and drawings. The examiner also conducts a search of available prior art references using search databases, including the Internet. Following the examiner's initial examination and search, the examiner will usually issue an objection to the application for failing to satisfy a formal requirement, or reject the claims as anticipated or obvious in view of the prior art discovered during the examiner's search.

In response to an Official Action, the applicant, typically through his or her attorney, can submit a formal response to address the rejections noted by the examiner and distinguish the claimed invention over the prior art. By distinguishing the claimed invention over the prior art, the applicant may amend the claims. Claim amendments are not required and may be particularly unnecessary when an examiner misinterprets a reference, or improperly combines references to support a rejection.

After filing a response to an Official Action, the examiner considers the arguments or amendments and makes a determination as to whether to issue a subsequent Official Action to allow the application. If the examiner issues another

Official Action, the Applicant will be given opportunity to respond. There is no limit on the number of Official Actions that can issue in the patent application process, although after a first action, examiners will usually issue a final office action which can have the effect of closing prosecution. If prosecution is closed in an application, an applicant can file a Request for Continued Examination (RCE) along with a response and the USPTO official fee. If the examiner decides to allow the application, the applicant will receive a Notice of Allowance, which will have a set period for the applicant to pay a fee in order to have the application officially issued as a patent.

1.5.6 Continuing Applications

United States patent law allows applicants to file continuing patent applications claiming the benefit of the disclosure and filing date of an earlier-filed co-pending application ("parent application"). The parent application does not have to be the first or earliest filed application in a chain of continuing applications; it just has to be a related application that is co-pending at the time of filing. Continuing applications must share at least one common inventor with the parent application, make a specific claim of priority to the parent application, and be filed while the parent application is co-pending. Although continuing applications claim the benefit of the earlier filed "parent" application, they are newly-filed applications that restart the examination process.

There are three types of continuing patent applications recognized in U.S. patent practice: (1) continuation applications; (2) continuation-in-part applications; and (3) divisional applications.

Continuation applications have the same specification as the parent application but with different claims. Continuation applications are useful to: (1) claim subject matter that was disclosed but not fully claimed in the parent application; (2) seek broader, narrower, or different claim coverage; (3) present new arguments in support of allowance of the application after a final rejection is received or prosecution is closed[4]; or (4) keep an application pending to capture developments not specifically addressed by any of the issued claims.

Divisional applications can be filed in response to an Office Action from the USPTO which states that the claims of the parent application are directed to two or more distinct inventions (e.g., claims to a product and claims to a method of making a product can be considered distinct inventions).

A continuation-in-part application ("CIP") is a later-filed application that repeats some substantial portion, if not all, of the parent application's disclosure,

---

[4] A Request for Continued Examination (RCE) under 37 C.F.R. § 1.114 can be filed upon payment of the requisite fee to present new arguments or claims in an application after a final rejection as an alternative to filing a continuation application.

and, generally, adds new subject matter not disclosed in the parent application. Claimed subject matter that is supported by the parent application is entitled to the effective filing date of the parent application. Claimed subject matter that is not supported by the parent application has the filing date of the CIP. Generally, CIPs claim new or related embodiments of an invention not disclosed in the parent application while effectively maintaining the filing date of the parent application for all originally disclosed subject matter. Although the applicant always has the option of filing a new application for this new subject matter, the priority claim for a CIP application may prevent the parent application itself from being cited to reject any original subject matter from the parent application that is claimed in the CIP.

Example: Dr. Fuse files a patent application describing a smartphone product as being a smartphone with a touch screen that is capable of being locked and unlocked by user-specific inputs, such as the user's fingerprint. Assuming that the smartphone with fingerprint recognition and the smartphone with voice recognition have a common composition, Dr. Fuse may be able to file a continuation-in-part (CIP) application to describe and claim the new embodiment to a smartphone product that utilizes voice recognition inputs. By filing a CIP application, the priority date for all subject matter in the CIP that overlaps with the original application disclosure will have the original application's filing date. All new subject matter will be entitled to the CIP application's filing date. Such a continuation-in-part application must be filed while Dr. Fuse's original patent application or an application claiming priority thereto is still pending (i.e., not an issued patent).

1.6 International Patent Rights

Rules for obtaining a patent differ from country to country. Patent protection in other countries requires international filings, usually with each country's patent office. Most countries permit applicants a non-extendible period of 1 year from the date of filing a U.S. patent application in which to file their patent application. In most countries, if a foreign patent application is filed within this 1 year period and claims priority to a U.S. patent application, the U.S. patent application filing date is the applicable priority date of the application.

The U.S. and approximately 120 other countries abide by the Patent Cooperation Treaty (PCT) that permits patent applicants to file international patent applications, also know as PCT applications. A PCT application is similar to a U.S. provisional application in that it preserves priority and never issues as a patent. Within 30 months from the PCT priority date, the applicant must file individual patent applications in all countries in which examination is desired (*i.e.*, PCT applications provide an additional 18 months time to file foreign applications beyond the typical 1 year period for filing priority foreign applications). Filing a PCT application can be advantageous in the following respects:

(1) if an applicant is interested in filing a patent application in numerous countries, a PCT application permits the applicant to have the benefit of a PCT patent examiner's prior art search and results before incurring the expense of filing numerous patent applications;

(2) a PCT application gives an applicant additional time (30 months from the PCT filing date) to delay the expenses associated with applying for patent protection in individual countries; and

(3) many countries give credence to a PCT examiner's examination search and opinion on patentability, which can limit the costs of prosecuting a patent application in individual countries.

Example: Dr. Fuse assigns his patent to the company that employs him. The company wants to pursue U.S. and foreign patent protection for the touch screen smartphone product. The company would like to file its patent applications as soon as possible, but is unsure as to how successful the product will be and is hesitant to spend too much on international patent protection.

If the company files a PCT application, it will have up to 30 months to determine in which countries to pursue protection. This will give the company additional time to evaluate the commercial success of the product and target select foreign markets. A PCT application will also give the company the benefit of a single examination, which can assist its determination of how much to invest in both U.S. and international patent protection.

Chapter 2
Trade Secret Protection

Trade secret law provides a mechanism for protecting proprietary and sensitive business information. A trade secret, by definition, is information that has economic value and is secret. There are no formal application requirements to obtain a trade secret. Unlike patents, there are no statutory requirements that a trade secret be novel, useful, non-obvious, and there is no examination process. Trade secret protection arises once the appropriate steps are taken to create and maintain a valid trade secret. Trade secrets are not subject to a predefined term, and can be maintained for an indefinite period of time.

2.1 What is a Trade Secret?

Unlike patent law, which has its roots firmly grounded in federal constitutional and statutory law, trade secret law is a state law doctrine that developed out of the common law doctrine of unfair competition and unfair business practices. Until passage of the Uniform Trade Secrets Act (UTSA) in 1985, trade secret law varied significantly from state to state. The UTSA is a model law that provides a uniform definition of trade secrets and misappropriation, and 45 states, the U.S. Virgin Islands, and the District of Columbia, have adopted it.

The UTSA defines a trade secret as "information, including a formula, pattern, compilation, program, device, method, technique, or process, that: (i) derives independent economic value, actual or potential, from not being generally known, and not being readily ascertainable by proper means by, other persons who can obtain economic value from its disclosure or use, and (ii) is the subject of efforts that are reasonable under the circumstances to maintain its secrecy." This broad definition maintains the common law that nearly any type of business information can qualify as a trade secret. Thus, information that is not otherwise patentable can

G. B. Halt, Jr. et al., *Intellectual Property in Consumer Electronics,*
Software and Technology Startups, DOI: 10.1007/978-1-4614-7912-3_2,
© Springer Science+Business Media New York 2014

be trade secret. Examples of information that can be protected by trade secret include:

• Computer programs	• Customer lists
• Client identities	• Vendors
• Product pricing	• Market analysis and strategies
• Manufacturing processes	• Formulas
• Technical information	• Product testing results
• Drawings	• Prototypes
• Strategic plans	• Company manuals
• Schematics	• Product ingredients
• Financial statements	• Employee records and salaries

Because information of nearly any type of subject matter can qualify as a trade secret, the UTSA definition of a trade secret focuses on: (1) the economic value of the trade secret, (2) whether the trade secret is generally known or readily ascertainable; and (3) the efforts taken to maintain secrecy. The "economic value" requirement under the UTSA refers to whether a competitor would obtain an economic benefit if the trade secret information became readily accessible. "Economic value" can be shown by the time and effort utilized in creating the trade secret, or by showing that a third party would have to spend time and effort in creating the same trade secret.

The second requirement for a trade secret under the UTSA is that the information cannot be "generally known or readily ascertainable." This means that the information cannot be already known to the public or by competitors. Whether a trade secret is "generally known or readily ascertainable" is a factual inquiry that depends on the amount of time, effort, and money required to independently produce the trade secret, or to reverse engineer the trade secret. Information cannot be protected by a trade secret if it can be discovered by examining a commercially available product that incorporates the information. If the trade secret is hidden in a commercially available product, then the trade secret can be maintained. A trade secret that consists of the amounts and ratios of individual ingredients in a product or code embedded in a software program is not lost just because the product becomes public availability.

Published information, such as that disclosed in a book, magazine, trade publication, website, or other media, cannot be maintained as a trade secret because it is "generally known" and readily ascertainable. This can be particularly important when deciding whether to keep information as trade secret or to pursue patent protection for that information. Anything disclosed in a patent or published patent application is generally known and readily ascertainable and cannot be protected as a trade secret.

The final and often most important criterion for a trade secret under the UTSA is that reasonable efforts must be taken to maintain secrecy of the information.

Maintaining secrecy of a trade secret is viewed under a reasonable standard which does not require absolute secrecy. A court considers several factual inquiries when considering reasonable secrecy:

- Whether employees have/have not executed confidentiality or non-disclosure agreements;
- Whether the company's confidentiality policy is memorialized in writing;
- Whether access to the trade secret is been limited to essential employees/ contractors;
- Whether employees who are privy to the trade secret are aware that it is to be maintained as a trade secret;
- Whether the information is kept in a restricted area such as a locked file, within security encrypted software, in a restricted location within a physical plant, etc.;
- Whether documents containing information that is trade secret are properly labeled; and
- Whether the company actively screens employee publications, presentations, etc. for disclosure of trade secret information.

In addition to these factors, it is important that the owner of the trade secret take steps to enforce secrecy of the information. Mere intent to keep information secret, without affirmative acts, is typically insufficient to maintain a trade secret.

2.2 Examples of Trade Secrets: Google's Search Engine Algorithm, WD-40 and Auto-Tune Software

- **Google's Search Engine Algorithm**: Google has become a staple in the lives of many individuals. It has become an invaluable tool for work, school and leisure. The search algorithm used in the famous search engine is a tightly guarded trade secret. PageRank is the proprietary search algorithm that fuels the Google search engine. Named after its creator and one of Google's founders, Larry Page (Google's other founder is Sergey Brin), PageRank organizes search results based on how frequently other sites link to the page, the website's content, popularity of the site and its location on the Internet. The PageRank algorithm was developed by Page and Brin while they were attending Stanford University. Since their invention was created with the use of Stanford's resources, the IP rights to the search engine algorithm were assigned to the university, who patented the idea behind the search engine algorithm, but licensed the right to use the algorithm back to Google on exclusive terms.
- **Water Displacement-40 (WD-40)**: The ubiquitous multi-purpose petroleum-based product is primarily used to lubricate, protect and displace moisture. WD-40 can be found in nearly 80 % of American households. The formula for WD-40 was perfected by Norm Larsen in 1953 on the 40th attempt to create a compound to prevent rockets from rusting. The multi-purpose spray formulation

was never patented and remains a well kept trade secret. The formula is written down and has been stored in a bank vault for over 50 years, being taken out of the vault only twice since then.

- **The Demise of the Auto-Tune Software Trade Secret**: Auto-Tune is proprietary software used to correct pitch and alter vocal effects in music, and has become standard equipment found in virtually all recording studios. The flagrant and obvious use of Auto-Tune technology in popular music is readily apparent today—singers such as diva Cher, rapper T-Pain, and country stars Faith Hill and Tim McGraw have all perceptibly used the technology in their music. However, when Auto-Tune was first developed in 1996, it was only used for the subtle correction of off-key inaccuracies in singing. Utilizing Auto-Tune software to digitally fix a few off-key notes saved the costs associated with rerecording the track, so naturally, studio engineers kept the technology a secret until Auto-Tune came to the attention of American ears when it was used, and arguably abused, in Cher's 1998 song "Believe." As use of the Internet and social media gained popularity throughout the 2000's and access to open-source code and user-generated content became easier, the Auto-Tune trade secret was ultimately disclosed to the public.

2.3 Misappropriation of Trade Secrets

A trade secret owner has the right to prevent others from misappropriating the trade secret. The UTSA defines misappropriation of a trade secret as:

(i) acquisition of a trade secret of another by a person who knows or has reason to know that the trade secret was acquired by improper means; or

(ii) disclosure or use of a trade secret of another without express or implied consent by a person who

    (A) used improper means to acquire knowledge of the trade secret; or

    (B) at the time of disclosure or use knew or had reason to know that his knowledge of the trade secret was

    (I) derived from or through a person who has utilized improper means to acquire it;

    (II) acquired under circumstances giving rise to a duty to maintain its secrecy or limit its use; or

    (III) derived from or through a person who owed a duty to the person seeking relief to maintain its secrecy or limit its use; or

    (C) before a material change of his position, knew or had reason to know that it was a trade secret and that knowledge of it had been acquired by accident or mistake.

In summary, misappropriation is the improper acquisition, disclosure, or use of a trade secret. A trade secret can be misappropriated even if the misappropriating party is not identically duplicating the trade secret.

Trade secrets can be lost or stolen in a variety of ways. Theft, bribery, misrepresentation, or breach of a duty to maintain secrecy are common acts that trigger a trade secret loss. Violating a confidentiality or non-disclosure agreement or obtaining the trade secret from a third party that is bound by a duty of confidentiality can give rise to an action for misappropriation. For example, a common means by which trade secrets can be lost or stolen is typically through unhappy or former employees who use or disclose the trade secret information apart from the company.

When a company discloses its trade secret to others, such as employees, manufacturers, suppliers, consultants, etc., those disclosures should be made under a written duty of confidentiality. This is typically done by requiring the party to execute a confidentiality or non-disclosure agreement, by way of employment contract, or third party consulting or supplier agreement. If a party under a duty of confidentiality with the trade secret owner breaches that duty, the trade secret owner's enforcement effort will benefit from a written agreement that clearly recognizes the trade secret status of the information.

Example: There is a difference between gathering competitive intelligence on a competitor and corporate espionage. Competitive intelligence is gathering information on a competitor legally and ethically, while corporate espionage involves illegal activities such as tapping a competitor's telephone lines or stealing their garbage to learn secret information about the competitor. Abusing a position of power or special credentials at one's old company to provide information to a new employer is also a form of espionage. Theft of trade secrets is often a serious problem when employees leave one company and move to a competitor.

For instance, while working as a project engineer at Netgear Inc. in 2005, Suibin Zhang was offered a position with Broadcom Corporation. Netgear has close business relations with Marvell Semiconductor Inc., and Broadcom is one of Marvell Semiconductor Inc.'s leading competitors. Due to his position at Netgear, Zhang had access to Marvell's secure database information and accessed that information after accepting the position with Broadcom but before leaving Netgear. For nine days, Zhang downloaded confidential information belonging to Marvell, including trade secret information. Marvell's trade secret information was later loaded onto a Broadcom-issued laptop by Zhang. The FBI apprehended Mr. Zhang, and found the trade secret information in Zhang's possession. He was charged and convicted of theft of Marvell's trade secret information.

The UTSA identifies a number of remedies for misappropriation of trade secrets including injunctions, damages, and attorney's fees. The UTSA even permits recovery of both the actual loss created by the misappropriation and any unjust enrichment resulting from the misappropriation that is not included in the "actual loss" portion of the damages. If actual loss for the misappropriation is difficult to prove, the trade secret owner may seek a "reasonable royalty" as compensation for the misappropriation. If the acts resulting in the trade secret misappropriation are willful or malicious, the UTSA grants the court discretion to award attorney's fees to the trade secret owner.

2.4 Case Study: Groupon Ex-Employees Take Trade Secrets to Google

"Employee poaching" is a controversial business practice whereby companies are accused of recruiting employees from competitors so that they can gain access to the competitor's trade secrets or technical know-now. The competing company dangles a tantalizing carrot for the employee it seeks to poach—better compensation, for example—in an effort to get the employee to defect. To prevent employee poaching, companies will often try and put restrictions on their employees to protect confidential company information. It is common for employment agreements to include non-compete clauses or a covenant not to compete. These clauses are designed to prevent employees from taking up employment with competitors.

Groupon, the discount coupon deal-of-the-day social website, recently faced this problem. After Google developed and launched its competing business model, Google Offers, two of Groupon's sales executives resigned from Groupon in September 2011 to work for Google Offers, purportedly for more money. In October of 2011, Groupon brought suit, alleging that the two employees, Brian Hanna and Michael Nolan, took trade secrets to Google that they would inevitably disclose. Groupon also alleged that the two employees breached their employment contracts, which prohibit them from taking a position with a company that provides similar services to those provided by Groupon. Since Hanna and Nolan were employed with Groupon at the executive level, they also contracted not to disclose any of Groupon's confidential business information or trade secrets, including any information regarding deal structures, pricing models, market research and customer data which they had encountered while employed with Groupon. Also according to the contract, Hanna and Nolan had to refrain from working for a competitor or soliciting Groupon's customers for two years post-termination of employment.

A few months later, the two employees countersued Groupon alleging sham litigation—the litigation being a ruse for Groupon to actually extract valuable information through third-party subpoena about its fledgling competitor Google Offers. Since trade secrets derive their economic value primarily from being unknown to competitors, it is clear why Groupon would be concerned with protecting trade secret information from misappropriation due to employees leaving with confidential information. However, the two employees say that they were never privy to confidential information making the employment agreement noncompete clauses entirely unnecessary and therefore, unenforceable and invalid. They point out that Groupon's business operations center around selling coupons through the Internet, a business model that is not all that dissimilar from traditional coupon-clipping. As such, they argued that the information cannot be subject to trade secret protection.

How could Groupon have clearly conveyed to its employees that company trade secret information existed in the workplace? Perhaps Groupon could have implemented better trade secret protection strategies and avoided this litigation, at least as it pertains to making a determination of whether trade secret protection exists. For instance, if Groupon's employees were properly put on notice as to when they are accessing trade secret information in the workplace, there would be little doubt in the mind of the employee that the information is a Groupon trade secret. Interoffice documents and emails could have been marked as confidential, access to certain client information could have had restricted access, and regular reminders to staff about the importance of keeping trade secret information confidential, are all good methods for demonstrating the existence of trade secret information that Groupon could have implemented.

At the time of this book's publication, the litigation between Groupon and its former employees, Brian Hanna and Michael Nolan had not been resolved.

2.5 Independent Discovery and Reverse Engineering of Trade Secrets

Under patent law, a subsequent inventor can be liable even though the invention was developed completely independently and without knowledge of the patented invention. Under trade secret law, independent discovery and use of the trade secret is not a violation. Further, competitors often try to uncover and trade-off of one another's trade secrets by "reverse engineering" the trade secret; a legally acceptable practice. The comments to the UTSA state that "reverse engineering" is a proper means of discovering a trade secret and identify reverse engineering as "starting with the known product and working backward to find the method by which it was developed. The acquisition of the known product must, of course, also be by a fair and honest means, such as purchase of the item on the open market for reverse engineering to be lawful...." Thus, discovery of another's trade secret requires proper acquisition of the information and ethical business practices.

Example: Code Rebel is a remote-access software solutions company that allegedly downloaded a trial version of Aqua Connect's "ACTS" software and then reverse engineered the program code, which it then used to create its own iRAPP Terminal Server solution. According to Aqua Connect, Code Rebel misappropriated Aqua Connects trade secret information. Despite the fact that the trial version of the ACTS software required acceptance of a clickwrap End User Licensing Agreement to access the program, with one of the terms being that the end user will not reverse engineer the program, the U.S. District Court for the Central District of California held that reverse engineering software in violation of a form end-user license agreement, without more, does not rise to the level misappropriation of a trade secret.[1]

---

[1] Aqua Connect, Inc. v. Code Rebel, LLC et al., Case No. CV 11-5764-RSWL (C.D. Cal. 2012) (Lew, J.).

Chapter 3
Trademarks and Trade Dress

3.1 What is a Trademark?

Use of symbols or signatures to identify the source of goods has been around since people first started trading and selling goods such as pottery, weapons, and clothing thousands of years ago. The purpose of these marks, to indicate the product's source, has not changed to this day. What has changed, especially in the last one hundred years, is the protection afforded to trademarks. Currently, the United States protects trademarks under the Trademark or Lanham Act, state law, and common law.

Under the federal Lanham Act, a trademark is any word, name, symbol, device, or any combination thereof that is used to identify and distinguish goods or services of one source from those of another source. In short, a trademark indicates the source of the goods or services. The law also provides protection for other types of marks that are directed to different types of uses. Many of these different types of marks are common in the high tech electronics and computer software industries, and the below table notes some of the key features of these different types of marks.

Type of mark	Key features	Example
Service mark	Used to identify and distinguish the source of services	*Internet Explorer*[a]
House mark	A "house mark" generally refers to a trademark that is used in all facets of a company's business, including business cards, letter head, packaging, and advertising. Typically, a house mark is also used with a secondary mark or can be used as a primary trademark	*HP*[b]

(continued)

G. B. Halt, Jr. et al., *Intellectual Property in Consumer Electronics,*
Software and Technology Startups, DOI: 10.1007/978-1-4614-7912-3_3,
© Springer Science+Business Media New York 2014

(continued)

Type of mark	Key features	Example
Trade dress	Trade dress refers to the overall impression created by a product which can be comprised of any combination of shape, color, design and wording. If trade dress is functional it cannot be registered or protected. Product design trade dress is not registerable until there is secondary meaning	*Texas Instrument's*[c] use of the outline of the state of Texas on its product packaging
Collective mark	Service mark used by the members of a cooperative, an association, or other collective group or organization which indicates membership in a union, an association, or other organization	*The Institute of Electrical and Electronics Engineers (IEEE)*[d] indicating membership in a professional organization that is scientific and educational. The group works toward advancing the theory and practice of electrical, electronics, communications and computer engineering, as well as computer science
Certification mark	Mark used to certify regional or other geographic origin, material, mode of manufacture, quality, accuracy, or other characteristics of someone's goods or services, or that the work or labor on the goods or services was performed by members of a union or other organization	*Energy Star*[e] for an international standard for energy efficient consumer products
Trade names	Used to identify a business or vocation. Trade names that merely identify a business are not registrable under the Lanham Act for federal registration. A trade name can also be a trademark if *used* as a trademark to indicate source. For example, Ford Motor Company can be both a trademark and a trade name	*GE Money Bank*[f]

[a] U.S. Registration No. 2,277,112. Owner: Microsoft Corp.
[b] U.S. Registration No. 3,801,893. Owner: Hewlett-Packard Development Company.
[c] U.S. Registration No. 3,717,043. Owner: Texas Instruments, Inc.
[d] U.S. Registration No. 1,770,511. Owner: Institute of Electrical and Electronics Engineers, Inc.
[e] U.S. Registration No. 1,999,485. Owner: Environmental Protection Agency.
[f] U.S. Registration No. 3,225,522. Owner: General Electric Co.

It is common for people to lump all of the above terms together as trademarks or "brands."

A trademark normally consists of a word, logo or some combination of the two. A word mark can include known terms, abbreviations, something coined by the owner, or some combination of letters and numbers. A logo can be a design, stylized lettering, or a drawing of an object. However, there are other types of trademarks, including the following:

- Symbol (Apple's iconic Apple logo[1]);
- Shape (shape of an iPod[2]);
- Slogan (*"Just Do It"*[3]);
- Sound (NBC chime[4]);
- Color (3M yellow Post-it notes[5]).
- Scents (fruit-scented industrial lubricants[6]);
- Touch ("velvet textured" feel on bottle surface for wines[7]);
- Distinctive Packaging (black and white cow print pattern on Gateway computer box[8]); and
- Building Design (University of Texas's UT Tower—a clock tower with an observation deck overlooking the UT campus and the city of Austin, Texas[9])

3.2 Brand Selection and Development

In the high-tech electronics or computer industries, selecting the right trademark for a product can mean the difference between success and failure. Successful brand management balances legal and business considerations. The primary legal consideration is the selection of the strongest trademark possible. When considering the strength of a trademark, trademarks are ranked on a sliding scale of distinctiveness ranging from unprotectable to extremely protectable.

---

[1] U.S. Registration No. 3,078,580. Owner: Apple Computers, Inc.

[2] U.S. Registration No. 3,341,214. Owner: Apple, Inc.

[3] U.S. Registration Nos. 1,875,307; 1,931,937 and 1,817,919. Owner: Nike, Inc.

[4] U.S. Registration No. 0,916,522. Owner: NBC Universal Media, LLC.

[5] U.S. Registration Nos. 2,402,722; 2,371,084. Owner: 3M Company.

[6] U.S. Registration No. 2,463,044. Owner: Mike Mantel.

[7] U.S. Registration No. 3,155,702. Owner: American Wholesale Wine & Spirits, Inc.

[8] U.S. Registration No. 1,725,231. Owner: Acer Inc.

[9] U.S. Registration Nos. 1,230,438 and 3,148,092. Owner: The Board of Regents of The University of Texas System.

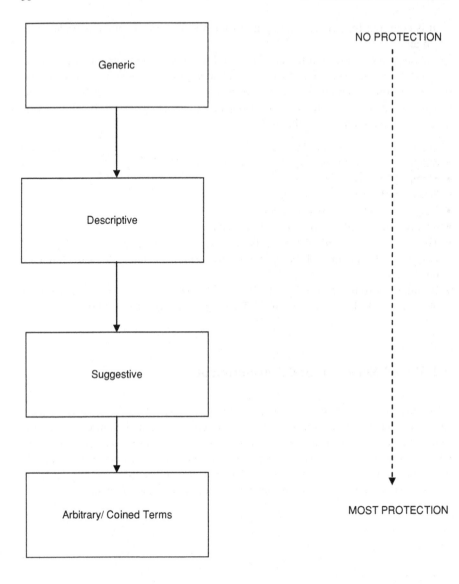

At one end of the scale are generic words. A generic word is a word that has come to be known as the common term for a class of goods or services. Generic designations are not registerable or protectable because they are incapable of functioning as a source indicator. A word can be inherently generic such as the word "monitor." Alternatively, a word can be "genericized" when the public associates the brand name as the product rather than the source. One example is "touch screen" for a type of monitor that is actually a touch screen monitor.

Next on the scale of protection are "merely descriptive" words that simply describe the product or convey an immediate idea of what the product does. One example is PATENTS.COM, which is merely descriptive of a feature of the good (computer software) for managing a database of records and for tracking the status of the records over the Internet.[10] Because descriptive marks simply describe a product, they are not protectable until they have acquired distinctiveness in the marketplace through their use. In other words, the trademark owner must show that a consumer primarily associates the mark with the product. In most cases, such distinctiveness takes years to acquire. The mark *"TEXAS INSTRUMENTS"*[11] for semiconductor and computer technology is a good example of a descriptive mark that became protectable after acquiring distinctiveness; it is descriptive because the company is located in Dallas, Texas and the company manufactures consumer electronics, i.e., electronic instruments. In addition, non-traditional trademarks such as sound, color and scent are not protected until they acquire secondary meaning.

On the business side, there is a strong temptation for a brand manager to select a trademark that is descriptive for the simple reason that a descriptive mark immediately provides the customer with information about the product. However, the downside to a descriptive mark is that the terms are normally used on a wide variety of goods, including the goods of competitors. If the company is able to obtain a descriptive trademark, it will be for a narrow set of goods. Descriptive marks can be problematic in the event that the product takes off in popularity and the trademark owner would like to parlay that success by expanding into other product areas.

Suggestive marks are next on the scale of protection and require some imagination, thought, or perception to come to a conclusion as to the exact nature of the goods. One fictional example is *DIGI-EXTERMINATOR* for a type of software debugging tool. Marks of this nature may be similar to a descriptive mark, but are registerable without a showing of acquired distinctiveness.

The most distinctive marks on the scale are marks that are either entirely coined, such as *EXXON*,[12] and cannot be found in any dictionary, or are arbitrary in the sense that they use common words in a way that is not expected, such as JELLIBEANS applied to skating rinks.[13]

By contrast, an arbitrary or fanciful mark is undoubtedly a strong mark. Since it bears no relationship to the goods or services, it usually takes more time and marketing efforts to create brand awareness among consumers. Once that awareness among consumers is present, the company can achieve and maintain extremely strong brand recognition. ROLEX for high-end wrist watches and jewelery[14] and DEBEERS for diamonds[15] are two well-known examples.

---

[10] In re Oppedahl & Larson LLP, 373 F.3d 1171 (Fed. Cir. 2004).

[11] U.S. Registration No. 3,717,043. Owner: Texas Instruments, Inc.

[12] U.S. Registration No. 3,736,430. Owner: Exxon Mobil Corp.

[13] Jellibeans Inc. v. Skating Clubs of Georgia, Inc., 212 U.S.P.Q. 170 (N.D. Ga. 1981).

[14] U.S. Registration No. 1,753,843. Owner: Rolex Watch U.S.A., Inc.

[15] U.S. Registration No. 3,510,098. Owner: De Beers.

3.3 Non-Protectable Subject Matter

Trademarks are only protectable if they are capable of distinguishing the goods or services of one owner from those of another. Therefore, generic marks cannot be protected and descriptive marks are only protectable with a specific showing that they have acquired distinctiveness. However, there are also additional types of marks that are not protectable.

(i) **Functional Marks.** Non-traditional marks such as colors and product designs are not protectable if they are, on the whole, functional. For example, the color yellow has been found to be functional when used in conjunction with safety products.

(ii) **Surname.** Marks that are primarily merely surnames are not protectable absent a showing of acquired distinctiveness. MCDONALDS and FORD are examples of surnames which have acquired distinctiveness.

(iii) **Immoral or Scandalous Marks.** The Lanham Act specifically bars immoral or scandalous matter from protection.

(iv) **Likelihood of Confusion.** Proposed use or registration of a trademark may be blocked by a prior trademark holder under common law or federal registration rights if the proposed use is identical or is likely to cause confusion with the existing trademark rights.

(v) **Geographic Descriptiveness.** Marks that are geographically descriptive cannot be registered until they acquire secondary meaning. An example of such a mark is TEXAS INSTRUMENTS.

(vi) **Names and Portraits.** Trademark protection is not available for marks that consist of the name, portrait or signature of a living person without their consent.

(vii) **Dilution.** A trademark that dilutes the distinctive quality of a famous mark is not registerable even if no likelihood of confusion is present.

3.4 Selecting a Trademark

In most cases, the process for selecting a trademark is not simple. Whether a company is a newly formed startup looking for a new house mark or a multi-national organization looking for a mark for a new niche consumer electronics product, a process for the selection of a trademark must be in place. The basic tenants of a typical selection process are discussed below.

3.4.1 Brainstorming Phase

During this phase, creative, marketing, and technical people develop a list of potential marks. If a company does not have the time or personnel to engage in a brainstorming session, it may hire an outside company that specializes in developing potential marks. Although such companies are normally good at what they do, they can be very expensive.

3.4.2 Narrowing Phase

During this phase, either a person or committee narrows the list of marks and those on the committee should consider some of the points already discussed, including:

- The commercial appeal of the proposed mark.
- The legal strength of the proposed marks. As noted, arbitrary or suggestive marks are typically more easily registered and defended.
- Whether to add a design feature or stylization.
- Whether competitors have similar marks.

Although the marks should be ranked in order of preference, each of the marks on the final list should be acceptable to the company as a trademark.

3.4.3 Knockout Phase

Before going through the expense of a trademark search (see next step), it is advisable to perform a "knockout" search to identify any obvious conflicts. A knockout search makes a determination as to whether the mark will be difficult to register (because, for example, the mark is scandalous, likely to be confused with some other mark, or the mark is already registered). A knockout search may be performed in-house or by an outside attorney on the available databases of registered marks to eliminate marks that may be difficult to register or conflict with a third party's marks.

3.4.4 Clearance Search

If a mark passes the knockout phase, a clearance search is conducted in the countries or jurisdictions of interest for the marks of interest. A clearance search assesses the availability of the mark for use or registration. These searches are typically conducted or overseen by either outside law firms or in-house personnel with trademark expertise. A search should include federal registrations, state registrations, common law marks, Internet domains, and websites.

When considering whether to conduct a full comprehensive search, consider the following points:

- How is the mark planned to be used?
- How widely will the mark be used?
- Will the mark be used on tooling for the product or just on advertisements?
- Will the mark be used on television?
- How important is the mark for the company?
- Will the mark be used in different countries or in different languages?

In setting a budget for these clearance searches, costs are directly proportional to the number of countries for which protection is sought. In most cases, the difficulty of clearing a mark increases the pool of potential problem marks increases. Therefore, a decision should be made early as to the intended countries in which the mark will be used. Although using an in-house search may be cost effective, it is generally recommended to have the clearance search performed outside of the company.

3.4.5 Obtaining an Opinion

The final stage in the selection process is whether to obtain an opinion as to the availability of a mark for use and registration in conjunction with a desired mark. Although obtaining an opinion based on the results of a search is not required, there are compelling reasons to obtain such an opinion.

First, it can be hard to assess search results. This is especially true in an industry where there are many competing products in other industries or the mark may be considered descriptive, (such as the high-tech electronics industry).

Second, although the clearance process can be expensive, the failure to clear a mark can be even more expensive. The ramifications of selecting a potentially confusing mark are serious. An infringer faces both the prospect of monetary damages and attorneys fees if found to infringe, the resulting embarrassment and loss of marketing momentum, and the finding of infringement would likely result in an injunction prohibiting use of the mark.

Third, obtaining a favorable opinion is strong evidence that there was no bad faith in using the mark and also shows that the brand manager is exercising due diligence. Both of these factors are important to a court when assessing willfulness and to others who may decide to second guess a trademark's selection.

3.5 Protecting the Mark

3.5.1 Common Law of Trademark

In the United States, unlike most countries, unregistered trademarks and names enjoy common law protection. This means that the party who adopts and uses a mark in a particular geographic territory is entitled to protection against a subsequent user who adopts the same or similar mark in that same territory. The concept of "territory" is a relatively nebulous and narrow concept that depends on the nature and extent of the use of the mark in a territory. For example, a business such as a restaurant that has limited advertisement or recognition in an area may only acquire trademark rights within a limited radius of its location under common law of trademarks. By contrast, a large company that advertises nationally and has sales throughout the United States may conceivably claim trademark rights throughout the entire United States.

Although reliance on common law rights may offer an initial costs savings, common law rights have several limitations set forth below:

- Limited to the particular territory where the mark was used;
- An innocent user who obtains a federal trademark registration may take over the rest of the country;
- Establishing common law rights is extremely fact sensitive. Accordingly, such rights can be difficult and extremely expensive to prove in court.

Example: Google announced in July 2012 that it will be entering the Internet service provider industry. Google will offer consumers Internet connections with speeds up to a billion bits per second via fiber-optic cables. The first installation of the fiber-optic infrastructure will be in Kansas City, Missouri, and installation will be done neighborhood by neighborhood. Google has playfully used the term "fiberhood" to describe the neighborhoods, which are signed up to receive the infrastructure.

Hypothetically, assume that Google does a trademark search of the word "fiberhood," but decides to delay filing a federal trademark application. Instead, Google immediately begins test marketing sales of its fiber-optic Internet in neighborhoods in and around Kansas City. After six months of better than expected sales and customer reviews, Google decides to launch the product nationally and seek federal trademark protection. Unbeknownst to Google, however, a small Seattle technology company innocently launches a fiber-optic Internet product under the same trademark, but has not registered its mark. As part of its marketing efforts the small Seattle technology company advertises in all the major newspapers in Washington and on local television and radio as well. Due to customer demand, all major electronic distributors in Washington carry the Seattle technology company's product. Under this scenario, even though Google is the senior party with first use, it is likely that the Seattle company will be able to continue to use its market in the limited market of Washington and perhaps in some areas of the surrounding states.

3.5.2 Federal Trademark Protection

Although a trademark owner can simply acquire geographic trademark rights through use of the mark, there are a number of advantages in filing for a federal trademark registration:

- **Right to use the ® symbol with all federally registered marks**. This symbol can have potent deterrent effects.
- **Provides constructive notice to the public of the claim to ownership of the mark**. Makes it much more difficult for a party to plead innocent infringement. Also, a basis for the United States Patent and Trademark Office ("USPTO") to reject confusingly similar marks. Common law marks cannot be cited by the USPTO to deny registration.
- **Confers nationwide priority of rights effective from the U.S. application filing date**. This may be the most important advantage. With this right, unlike with common law, a trademark owner does not have to prove use in a particular state or states(s) in order to claim trademark rights.
- **The legal presumption of the registrant's ownership of the mark, its validity, and the registrant's exclusive right to use the mark nationwide**. With this right the trademark owner once again does not have to prove rights in the mark, they are presumed.
- **Ability to bring an action concerning the mark in federal court and possible recovery of treble damages and attorney's fees**.
- **U.S. registration may serve as a basis to obtain registration in foreign countries without first using the mark**.
- **Ability to file the U.S. registration with the U.S. Customs Service to prevent importation of infringing foreign goods**.
- **Availability of incontestability status after 5 years of continuous use and registration**.

Although a trademark owner may gain some rights through use of a mark, it is recommended to file a trademark application as soon as possible and early in the development of the product.

The Federal Trademark Application Requirements

The requirements for filing a trademark application are relatively straightforward. In order for an application to receive a filing date, it must include: (1) the required filing fee for at least one class of goods or services; (2) the name of the applicant; (3) the name and address for the applicant or attorney for communication; (4) a clear drawing of mark to be registered; and (5) the identification of the goods and/or services that the mark will be used with.

The most complex part of filing an application is preparing the identification of goods and/or services for which trademark protection is being sought. The goal is to draft identification of goods and services as broadly as possible because the identification cannot later be expanded. The USPTO requires that the identification be specific and definite. Moreover, if use is claimed, it is important that the mark be used on all of the goods or services, and that there be a *bona fide* intent to use the mark if the application is filed on an intent to use basis. If these requirements are not met, any subsequent registration could be subject to cancellation for fraud on the USPTO.

There are four bases on which to register a mark:

1. Actual use in commerce;
2. Bona fide intention to use mark in commerce;
3. Foreign registration—this is only available to companies domiciled outside the United States; and
4. Under the Madrid Protocol—this is only available to companies domiciled outside the United States.[16]

In order for an applicant to claim use in commerce as basis for an application, the mark must have been in use as of the application date. The date of first use is the date on which the goods were first sold or transported in interstate commerce or the services first rendered anywhere in the world in an arm's length transaction. The date of first use in commerce is the date that the goods are either sold or transported in commerce such that they could be regulated by applicable laws. For services, the date of first use in commerce is the date the mark is first used or displayed in sales or advertising of services and the services are rendered in interstate commerce.

Once the trademark owner decides to adopt and use a trademark, an intent-to-use application should be filed as soon as possible in order to gain the advantage of constructive use date.

Examination

Following the filing of a trademark application, the USPTO assigns it to a trademark examining attorney who examines the mark to determine whether it is entitled to registration. The trademark examining attorney conducts a trademark search to determine if the mark is likely to cause confusion with any other mark on the Principal Register and reviews the application for compliance with the Trademark Act and USPTO Rules.

If for any reason the trademark examining attorney determines that the mark is not registerable for any reason, the trademark examining attorney will issue an Office Action that advises the applicant of all grounds of refusal and all matters that require further action. The applicant has 6 months to respond to the Office Action. This 6 month period runs from the mailing date of the Office Action.

---

[16] Foreign trademark protection is discussed below in this chapter.

Failure to fully respond to the Office Action within the statutory period results in the application becoming abandoned.

Once the applicant has had the opportunity to respond to all issues raised in the Office Action, the examining attorney issues a final communication that allows or finally rejects the application. After a negative final Office Action, the applicant may file a request for reconsideration and submit additional evidence and argument in order to persuade the examining attorney to withdraw the final Office Action. If the examiner fails to withdraw the final Office Action within 6 months, the Applicant must either meet every requirement of the Office Action or appeal to the Trademark Trial and Appeal Board in order to avoid abandonment of the application.

Following examination, if it appears that a mark is entitled to registration on the Principal Register and there are no outstanding requirements or refusals, the examining attorney will approve the mark for publication in the USPTO *Official Gazette*. If a third party does not file an opposition within thirty (30) days of publication or request a time extension to file an opposition, the application will proceed to registration.

For applications based on intention to use and where no Amendment to Allege Use has been filed before publication, the USPTO will issue a Notice of Allowance. The application will proceed to registration upon the filing of a Statement of Use. A Statement of Use or request for extension of time to file a Statement of Use must be filed within 6 months of the mailing date of the Notice of Allowance. The applicant may request an extension for a 6 month period without showing good cause. Thereafter, the applicant may receive an additional 6 month extension upon request and by a showing of good cause.

3.5.3 State Registration

State laws also provide for trademark registrations. However, given the many advantages to Federal Registration, there is little point to obtaining a state registration unless the trademark owner cannot establish use in interstate commerce (i.e., all of the applicants' use of the mark is within a single state), or has a specific legal need to take advantage of that state's trademark or anti-dilution remedies.

3.5.4 International Protection

Trademark protection is also available internationally. Unlike the U.S., most countries award trademark rights solely on a first-to-file basis. Therefore, it can be extremely important to consider the need to file a trademark application in other countries. A U.S. registration can form the basis of an application in a foreign country without the necessity of first having used the mark in that country. In most countries, a U.S. company can file a trademark application that claims the same U.S. application date if the trademark application is filed within 6 months of the U.S. filing date. However, foreign trademark protection is often very expensive and the costs multiply depending on the jurisdictions in which protection is sought. Accordingly, a company should develop a list of countries for which registration will be sought.

When considering which countries to pursue protection, the following factors should be evaluated.

- Serious consideration should be given to registering in those countries where the mark is used, or will be used in the near term.
- Consideration should also be given to countries where the company is planning on expanding in the next 3–5 years. This is especially true for large markets such as China, which uses a first-to-file system.
- Finally, if counterfeiting is a problem, consideration should be given to defensive filings in some of the key counterfeiting source nations such as Taiwan, China and Vietnam.

Once the decision is made, a trademark owner has several options for filing overseas. In many cases, the filing is done in each individual country, which requires hiring a trademark attorney in each country and paying filing fees in each country. However, there are some international treaties that allow a trademark owner to avoid some of these fees and costs.

Madrid Protocol

Foreign filing costs can be reduced by using the Madrid Protocol. The Madrid Protocol is an international trademark treaty which permits the owner of a "home country" registration to file an international application with its national trademark office that designates other member countries. The Madrid Protocol offers cost savings and increased efficiency for U.S. trademark holders. The International Trademark Association has aptly summarized the benefits of the Madrid Protocol as offering[17]:

- one application;
- in one place;
- with one set of documents;
- in one language;
- with one fee;
- resulting in one registration;
- with one number;

---

[17] International Trademark Association [1]

- and one renewal date;
- covering more than one country.[18]

The cost savings of registration through the Madrid Protocol are significant. Another advantage of the Madrid Protocol is the simplicity in filing application amendments. Without the Madrid Protocol, applications would have to be filed and prosecuted individually in every country in which the mark is registered. However, the Madrid Protocol simplifies this process at a reduced cost.

Other advantages of an International Registration under the Madrid Protocol include having priority of protection in all designated countries from the date of international registration, as opposed to the date of registration in the individual countries. Also, the Madrid Protocol limits the time a national office has to act once it receives a request for extension of a Madrid registration. If the office does not act to oppose protection during the allotted time, the registration is automatically granted.

By offering simultaneous registrations in the U.S. and foreign countries, the Madrid Protocol also reduces trademark piracy. Without the Madrid Protocol, individuals in foreign countries are often free to register a U.S. company's trademark, and attempt to sell the mark to the U.S. company at highly inflated prices. Registration under the Madrid Protocol reduces such piracy since all designated countries are given the same priority date.

Notwithstanding the cost savings and increased efficiency associated with the Madrid Protocol, there are some drawbacks for U.S. applicants. First, the Madrid Protocol requires that the scope of goods or services covered by the registration be limited to the home country's registration rules. U.S. applicants that seek registration through the Madrid Protocol will be prejudiced in this respect since the USPTO requires more detailed identification of goods and services than most other countries. Unlike some other countries, the USPTO will not accept registration of marks for broad classes of goods and services. Other countries, such as Egypt, allow for registration of a mark for a whole class of goods or services (i.e., clothing generally), without specifying a good or service within the class (i.e., t-shirts specifically). Therefore, U.S. companies may limit the scope of protection that could otherwise be obtained in other countries by filing an International Registration as opposed to filing individual national applications.

---

[18] The member countries to the Madrid Protocol include: Albania, Algeria, Antigua and Barbuda, Armenia, Australia, Austria, Azerbaijan, Bahrain, Belarus, Belgium, Bosnia and Herzegovina, Botswana, Bulgaria, China, Columbia, Croatia, Cuba, Cyprus, Czech Republic, Democratic People's Republic of Korea, Denmark, Egypt, Estonia, European Union, Finland, France, Georgia, Germany, Ghana, Greece, Hungary, Iceland, Iran, Ireland, Israel, Italy, Kazakhstan, Kenya, Kyrgyzstan, Latvia, Lesotho, Liberia, Liechtenstein, Lithuania, Luxembourg, Madagascar, Mexico, Monaco, Mongolia, Montenegro, Morocco, Mozambique, Namibia, New Zealand, Norway, Oman, Philippines, Poland, Portugal, Republic of Korea, Republic of Moldova, Romania, Russian Federation, San Marino, San Tome and Principe, Serbia, Sierra Leone, Singapore, Slovakia, Slovenia, Spain, Sudan, Swaziland, Sweden, Switzerland, Syrian Arab Republic, Tajikistan, The former Yugoslav Republic of Macedonia, Turkey, Turkmenistan, Ukraine, United Kingdom, United States of America, Uzbekistan, Viet Nam, and Zambia.

Another potential consequence of the Madrid Protocol for U.S. trademark owners is the limitation that the USPTO imposes on applicants to provide a statement of use or *bona fide* intent to use the mark in commerce before obtaining a filing date, and proof of use in commerce before a registration will issue. Most other countries do not require a similar statement of use or intent to use the mark in commerce, or proof of such use. Therefore, U.S. trademark applicants may be disadvantaged by their inability to "reserve" a trademark under USPTO procedure.

Trademark owners filing under the Madrid Protocol are also subject to "central attack." If a home country application or registration is cancelled or abandoned during the first 5 years of registration, whether completely or partially, the home country must notify World Intellectional Property Organization (WIPO). The International Registration then lapses with respect to all designated countries. This is particularly disadvantageous to U.S. trademark owners because there are usually more grounds for challenging registrations under U.S. law than in other countries.

The Madrid Protocol provides for a partial safeguard against "central attack" by providing a 3 month grace period for the owner of a cancelled registration to file national applications in designated countries that enjoy the same priority as the International Registration. This process, however, can be costly and time consuming.

Unlike national application systems, under the Madrid Protocol an assignment may only be recorded if the assignee is itself qualified to file a Madrid Protocol application. Although this only effects the recordation of the assignment and national laws will govern the legal effect of the assignment, member countries such as the U.S. have passed laws which make assignments to non-member citizens invalid. This assignment provision may be problematic in cases where a U.S. citizen or corporation wishes to assign registration(s) to a non-member citizen or corporation for tax or other purposes.

The final drawback of implementation of the Madrid Protocol for U.S. trademark owners is that the system is outside of the U.S. and the procedures can seem unfair for those used to U.S. filings. For example, the time period for responding to Office actions under the Madrid Protocol may be quite short due to the fact that an Office action is sent from the national office to WIPO and WIPO sends it to the trademark owner. Moreover, many Madrid Protocol countries do not send a Registration Certificate or other notice once the registration issues. Therefore, the trademark owner is left in the uncertain position of not knowing for sure if the registration has actually issued.

European Community Trademark

The European Community system offers a trademark system that allows for registration of a trademark in all of the member countries for one application filing fee.[19] If the trademark owner intends to use the mark in more than two European

---

[19] Member countries include: Austria, Benelux (Belgium, the Netherlands and Luxembourg), Bulgaria, Cyprus, the Czech Republic, Denmark, Estonia, Finland, France, Germany, Greece,

Community member countries, it is typically more cost effective to file for a Community Registration. Although a Community Registration can be canceled for 5 years of non-use, use in one country is enough to satisfy the use requirement.

3.6 Maintaining Trademark Rights

Trademark rights can be a company's most valuable asset. Like any asset, these rights need to be protected and maintained. Trademark rights are lost when the mark no longer acts as an identifier of the goods or services. This can occur through abandonment from non-use or through a course of conduct including acts of commission or omission which allow a mark to become the generic name of goods or services or lose significance as a mark.

3.6.1 Maintaining Federal Registrations

Section 8 of the Lanham Act requires that an affidavit or declaration verifying continued use in commerce or excusable nonuse due to special circumstances be filed with the USPTO between the fifth and sixth anniversary and on or between the ninth and tenth anniversary date of the registration. This requirement applies to all registrations. Failure to meet this requirement will result in cancellation of a registration.

The duration of a trademark registration has varied over the years. Since November 16, 1989, however, all registrations that have been issued or renewed after that date have only a 10 years term. Therefore, in order to maintain a federal trademark registration, a renewal application must be filed on or between the ninth and tenth anniversary date of the registration.

3.6.2 Licensing

A trademark owner can license its trademark to a third party. However, a trademark owner must reserve the power to exercise quality control over the nature and quantity of the goods and services in a license. If the trademark owner fails to exercise such quality control, the license is considered a "naked license" and the mark may be abandoned.

(Footnote 19 continued)
Hungary, Ireland, Italy, Latvia, Lithuania, Malta, Poland, Portugal, Romania, the Slovak Republic, Slovenia, Spain, Sweden and the United Kingdom.

3.6.3 Assignments

A trademark may only be assigned to another party with the goodwill of the business in which the mark is used. Failure to meet this requirement can result in trademark becoming void. Similarly, intent to use applications cannot be assigned prior to the filing of a Statement of Use or Amendment to Allege use unless the assignment is to the successor to the ongoing and existing business of the applicant or to the portion to which the mark pertains.

3.6.4 Genericide

There are many examples where once valuable trademarks have become generic (i.e., the trademarks have ceased to function as a source indicator and have become the name of a particular type of a product). The following terms were at one time a company's trademark:

- ALE HOUSE (for restaurant and bar services)[20]
- ASPIRIN (for acetyl salicylic acid pain reliever)[21]
- CELLOPHANE (for transparent cellulose sheets and films)[22]
- COLA (for soft drink)[23]
- CRAB HOUSE (for seafood restaurant)[24]
- CUBE STEAK (for steaks)[25]
- DRY ICE (for the solid form of carbon dioxide)[26]
- ESCALATOR (for moving stairs)[27]
- FONTINA (for cheese)[28]
- HOAGIE (for a sandwich)[29]
- HONEY BROWN (for a brown ale made with honey)[30]
- JUJUBES (for gum candy)[31]

---

[20] Ale House Management, Inc. v. Raleigh Ale House, Inc., 205 F.3d 137 (4th Cir. 2000).

[21] Bayer Co. v. United Drug Co., 272 F. 505 (S.D.N.Y. 1921).

[22] DuPont Cellophane Co., Inc. v. Waxed Products Co., Inc., 85 F.2d 75 (2d Cir. 1936).

[23] Dixi-Cola Laboratories v. Coca-Cola Co., 117 F.2d 352 (4th Cir. 1941).

[24] Hunt Masters, Inc. v. Landry's Seafood Rest., Inc., 240 F.3d 251 (4th Cir. 2001).

[25] Spang v. Marzall, 104 F. Supp. 126 (D.D.C. 1952).

[26] Dryice Corp. of Am. v. Louisiana Dry Ice Corp., 54 F.2d 882 (5th Cir. 1932).

[27] Haughton Elevator Co. v. Seeberger, 85 U.S.P.Q. 80 (Comm'r. Pat. 1950).

[28] *In re* Cooperativa Produttori Latte E Fontina Valle D'Acosta, 230 USPQ 131 (TTAB 1986).

[29] Raizk v. Southland Corp., 121 Ariz. 497, 499 (Ariz. Ct. App. 1978).

[30] Genesee Brewing v. Stroh Brewing, 124 F.3d 137 (2d Cir. 1997).

[31] Henry Heide, Inc. v. George Ziegler Co., 354 F.2d 574 (7th Cir. 1965).

- LIGHT BEER (for beer with fewer calories)[32]
- MONTESSORI (for educational services)[33]
- MURPHY BED (for folding bed)[34]
- SOFTCHEWS (for chewable medical pills)[35]
- SUPER GLUE (for glue)[36]
- SURGICENTER (for surgical center)[37]
- TOUCH TONE (for phones)[38]
- TRAMPOLINE (for jumping and gymnastic equipment)[39]
- ZIP CODE (for mailing designations)[40]

These terms are now generic and available for use by all. If a trademark owner wishes to avoid that result, it is essential that all those associated with a trademark understand the requirements. The following tips can help protect the value of a trademark.

- Do not trademark the name of your product. If you cannot provide a generic name of your product without referring to your brand, then the trademark may become generic.
- When using a trademark in a sentence, always use the trademark as an adjective. If this rule is not followed, the public may come to see the trademark as the generic name of the product or service which is what happened to former trademarks such as escalator, cellophane, and kerosene. Use of the word "brand" also can help emphasize that a term is a mark and not a generic descriptor, for example: Loctite® brand adhesive products. There is one caveat to this rule—many companies use their trade names as trademarks. In those instances where the trade name is being used, it is a proper noun, not an adjective. For example, one of Ford's marketing slogans incorporates its trade name into the slogan as a proper noun: "Have you driven a Ford lately?"
- In a sentence, the trademark should be set apart from the text in some fashion. This can be accomplished by using ALL CAPITAL LETTERS; **bold face type**; Initial Capital Letter; *italics*; or through the use of a unique font.

---

[32] Miller Brewing Co. v. G. Heileman Brewing Co., Inc., 561 F. 2d 75, 79-81 (7th Cir. 1977).

[33] American Montessori Society, Inc. v. Association Montessori Internationale, 155 U.S.P.Q. 591 (1967).

[34] Murphy Door Bed Co., Inc. v. Interior Sleep Systems, Inc., 874 F.2d 95, 101 (2d Cir. 1989).

[35] Novartis Consumer Health, Inc. v. McNeil-PPC, Inc., 1999 U.S. Dist. LEXIS 20981 (D.N.J. 1999).

[36] Loctite Corp. v. National Starch & Chem. Co., 516 F.Supp. 190 (S.D.N.Y. 1981).

[37] Surgicenters of America, Inc. v. Medical Dental Surgeries, Co., 601 F.2d 1011 (9th Cir. 1979).

[38] U.S. Registration No. 0,737,312. Owner: AT&T Co. Canceled March 13, 1984.

[39] Nissen Trampoline Co. v. American Trampoline Co., 193 F. Supp. 745 (S.D. Iowa 1961).

[40] U.S. Registration No. 1,042,499. Owner: United States Postal Service. Expired: April 7, 1997.

- Monitor Use of Your Mark. If you see your mark starting to appear in all lower cases in publications, this is a danger sign. Steps should be taken to send notifications to advertisers using the mark in that way.
- Police Misuse. If your trademark is found in a dictionary, whether intentional or unintentional, it is strong evidence that the trademark is generic. Corrective action in the form of a letter to the publisher should be taken immediately. Likewise, if you see your mark appear in lower case letters, this is a danger sign.
- Use the brand in a consistent manner. Not only is recognition of the mark enhanced through consistent use, inconsistent use may confuse consumers, dilute the distinctiveness of the mark, and lead to abandonment of the mark.
- Use the appropriate trademark designation. In the United States the symbol TM can be used to identify an unregistered trademark, SM can be used to identify an unregistered service mark, and ® can be used to identify a registered trademark or service mark. Many foreign countries use similar terms. Local laws should be consulted because many countries require proper use of the symbol in their country in order to collect damages for infringement. Such symbols may not be the same as those accepted in the United States.
- Develop a trademark usage manual. All companies should have a manual to advise employees and others on the proper use of the company's trademarks.
- Non-Use. A trademark owner must use the mark to maintain it. Three years of non-use results in a presumption of abandonment.
- Infringement. The key function of a trademark is as a source identifier. If the same or similar trademark is used by more than one company on the same or related goods, the mark may cease to be a source identifier. Therefore, it is important for a company to police third party usage of its marks and take appropriate action ranging from cease and desist letters to legal action.

3.7 Case Study: Google Polices the Use of Google Trademark, Protecting Against Genericide of the Mark

When a trademarked product is a market dominator to such an extent that the public identifies the company's trademark with all other similar products, the trademark runs the risk of suffering from genericide and losing its trademark protection. Google actively polices the use of its famous GOOGLE mark, in an effort to prevent the word from becoming synonymous with the term "Internet searching."

Google's first exhibited its tenacity for protecting its mark in 2003 with Google's response to Word Spy's inclusion of the word "google" in its popular online dictionary. Word Spy had noted the definition of "google" as: "to search for information on the web, particularly by using the Google search engine; to search the web for information related to a potential new girlfriend or boyfriend." Google's IP counsel sent a letter to Word Spy expressing concern over the definition, describing how it wanted to ensure that when people use the word "Google," that they are not referring to Internet searching in general, but rather are referring to the services provided by the Google company. Word Spy responded by including Google's trademark information, which ameliorated the concern of Google's IP counsel.

Three years later, in 2006, the word "google" was officially sanctioned as part of the English language by its inclusion in the Merriam-Webster's Collegiate Dictionary. The definition was crafted with Google's sensitivity about genericide of the mark in mind. "Google" is defined as a transitive verb meaning "to use the Google search engine to obtain information about (as a person) on the World Wide Web." The definition specifically identifies use of Google's search engine to perform an Internet search.

Earlier that same year, Pontiac aired a commercial in which it encouraged viewers to Google "Pontiac" to find out more about its automobiles. Pontiac had sought permission from Google to use its mark in its Pontiac commercial, in conjunction with images of Google's search engine being used to search for more information about Pontiac on the web. Despite the appearance that this was a generic use of Google's mark in commercial advertising, two factors make it policed use of the mark. Firstly, Pontiac had permission to use the mark. Secondly, the commercial featured images of the Google search engine, thus cementing in the minds of consumers that "Google" specifically refers to use of the Google search engine.

In 2007, and again in 2012, Google was faced with lawsuits seeking cancellation of the Google Trademark. In both cases, the plaintiff's sought to cancel the Google mark alleging that "Google" had passed into common usage. While the 2012 case is pending at the time this book is being written and the outcome is yet to be known, the 2007 plaintiff did not succeed. So long as consumers readily associate the term "Google" with "searching the Internet via the Google search engine" it is unlikely that the mark has passed into generic usage. It is unlikely that the 2012 plaintiff will succeed since people do not "Google" on the Yahoo! or Bing search engines.

Google is very diligent in the policing of its mark. Google's counsel is constantly monitoring the use of its mark by others and regularly sends out cease and desist letters to those it determines are abusing Google's trademark rights. So long as consumers associate the GOOGLE mark with Google the company, Google need not be concerned that genericide has taken protection away from its mark.

Reference

1. International Trademark Association, The Madrid Protocol: Impact of U.S. Adherence on Trademark Law and Practice, at p. 1 (Revised April 2003).

Chapter 4
Copyrights

4.1 Copyrightable Subject Matter and Scope of Protection

The federal Copyright Act protects authors' creative works. "Copyright protection subsists... in original works of authorship... fixed in a tangible medium of expression...."[1] Copyright automatically arises upon creation and fixation of an original work. In the high technology electronics industry, fixation of an original work may include advertising and marketing materials or packaging, schematic circuit diagrams, blueprints and software.

There are limitations on the scope of copyright protection. First, while registration is not required to secure copyright protection, it is a prerequisite to litigating a copyright claim and is desirable to preserve the remedies for attorney's fees and statutory damages. Second, "[i]n no case does copyright protection for an original work of authorship extend to any idea, procedure, process, system, method of operation, concept, principle, or discovery...."[2] Third, where a "useful article" is concerned, copyright protection is only afforded to the extent the original creative expression is physically or conceptually separable from the utility of the useful article. Finally, certain subject matter is not subject to copyright.

4.1.1 Circuits and Chips as Copyrightable Subject Matter: Mask Works

Copyright protection of a specific schematic drawing or integrated circuit layout may be granted; however, the underlying concept embodied by the schematic cannot be protected by copyright. For example, a creative schematic drawing of a circuit may be copyrighted regarding the orientation and arrangement of the electronic components in the circuit. The copyright protection does not prevent

---

[1] 17 U.S.C. § 102(a).
[2] 17 U.S.C. § 102(b).

G. B. Halt, Jr. et al., *Intellectual Property in Consumer Electronics,*
Software and Technology Startups, DOI: 10.1007/978-1-4614-7912-3_4,
© Springer Science+Business Media New York 2014

others from using the same components in a different configuration. Another person could copy the underlying concept of the diagram, resulting in a circuit that is functionally the same.

Mask works fixed in semiconductor chip products are also protectable under the Semiconductor Chip Protection Act (SCPA).[3] The SCPA is meant to protect the topography of semiconductor chips. A mask work is defined as a series of related images, however fixed or encoded:

(1) having or representing the predetermined three-dimensional pattern of metallic, insulating, or semiconductor material present or removed from the layers of a semiconductor chip product; and
(2) in which series the relation of the images to one another is that each image has the pattern of the surface of one form of the semiconductor chip product.

The SCPA defines a semiconductor chip product as the final or intermediate form of any product:

(1) intended to perform electronic circuitry functions; and
(2) having two or more layers of metallic, insulating, or semiconductor material, deposited or otherwise placed on or etched away or otherwise removed from a piece of semiconductor material in accordance with a predetermined pattern.

Chip protection can be acquired through the U.S. Copyright Office under the SCPA by filing a "mask work" application in conjunction with the appropriate filing fee. By registering the mask work, the registrant may affix a notice to the mask work to indicate ownership—notice is optional. The SCPA requires the form of the notice to be:

(1) the words "mask work," the symbol *M* or the symbol M; and
(2) the name of the owner(s) of the rights in the mask work or an abbreviation by which the name is recognized or generally known.

The duration of protection is 10 years, during which the mask work owner has the exclusive right:

(1) to reproduce the mask work by optical, electronic, or any other means;
(2) to import or distribute a semiconductor chip product in which the mask work is embodied;
(3) to induce or knowingly to cause another person to do any of the acts described in number (1) and number (2).

Protection begins on the date the mask work is registered with the Copyright Office, or the date the mask work is first commercially exploited anywhere in the world, whichever occurs first. Legal actions regarding infringement of a mask work can be brought in federal court.

---

[3] 17 U.S.C. § 901, *et seq.*

4.1.2 Software as Copyrightable Subject Matter

Software and computer programs are considered literary works for the purposes of copyright protection. Copyright protection only protects expressive elements of the program code and cannot be used to protect the functional aspects of the program. For example, methods of operation within the software, such as menu commands, for the most part are not copyrightable unless they are novel or consist of new and original artistic elements. Graphical User Interfaces (GUIs) elements are generally copyrightable. Copyright can also be afforded to both the source code and the object code that make up a program. The author may choose if he or she would like to seek copyright protection for the entirety of the program code, or only a portion of the code. Sometimes, graphic displays are used to illustrate the underlying operation of a program's function. Like schematics, these graphic representations, such as a flow chart or diagram, can be protected under a separate copyright in addition to the software code.

4.1.3 Advertising, Marketing Materials and Packaging

Original authorship is often embodied in advertising, marketing materials and packaging. The primary issues of concern in these areas are copyright ownership, avoiding infringement, and preserving remedies through registration.

Through registration, a copyright owner preserves its ability to recover attorney's fees and statutory damages, remedies that are not available if infringement commences before registration. The registration process also compels an examination of authorship/ownership issues, as well as necessitating a review of whether any preexisting material is embodied in a work which may give rise to a need for obtaining permission for use of such material. Pursuing registration of the copyright in a commercial should raise the issue of identifying all of the "authors" of the commercial and the use of preexisting material, if any. This will normally bring to light the need for appropriate assignments and clearances.

Example: Packaging, logos and labels can be protected under copyright as artistic works. An example of a copyrighted logo can be found in the fashion industry. The Louis Vuitton Multicolore Monogram graphic design used on its line of hand bags is protected by copyright and Louis Vuitton often enforces the copyright on that design.[4] Similarly, courts have held that the

---

[4] *See,* e.g., Louis Vuitton Malletier, S.A. v. Mido Trading, Inc., CV 08-04405 DDP (C.D. Cal. April 22, 2010); Louis Vuitton Malletier, S.A. v. Akanoc Solutions, Inc., 2009 WL 1636914 (N.D. Cal. 2009).

> Pledge label used on Pledge furniture cleaning products is protected by copyright.[5]

To the extent that packaging embodies creative expression, packaging is protectable under copyright. This can extend to the shape of packaging such as a Mickey Mouse-shaped popsicle, or a sculpted perfume bottle stopper. Containers are, however, generally useful articles that would be more appropriately protected by patents. For example, the packaging used by Apple to package iPods is patented.[6] In some instances, distinct containers may also be protected under trademark registrations such as the Tiffany & Co.'s signature Tiffany Blue Box.[7]

4.1.4 Training Manuals, User Manuals and the Like

Training materials, user manuals, company brochures, client alerts, instruction guides and company handbooks can be copyrightable if they demonstrate the requisite minimal degree of creativity for copyright protection. These types of documents qualify for copyright protection as "literary works." Copyright in such materials is limited to the description of the processes describe in the materials—and does not protect the process itself.[8] For example, if a training manual provides a step by step description for how an employee is to procure a reimbursement from the employer company, the written description, along with any flow charts or diagrams, is copyrightable. However, the process itself cannot be protected. Other companies could use the same process to reimburse employees.

Employer companies should be mindful of the need for assignment of IP rights from those employees who produce these kinds of documents are part of their employment. If an employee produces a training or user manual as a byproduct of his or her job, the employer company should make sure that the IP rights to the manual are assigned to the company.

4.1.5 Secret and Other Materials Including Software

In many instances, a business may develop software that is incorporated into high-tech consumer electronics products. To the extent such works contain original

---

[5] S.C. Johnson & Son, Inc. v. Turtle Wax, Inc., 1989 WL 134802 (N.D. Ill. 1989); *see also* Drop Dead Co. v. S.C. Johnson & Son, Inc., 326 F.2d 87 (9th Cir. 1963).

[6] U.S. Patent No. 7,878,326.

[7] U.S. Registration No. 2,359,351. Owner: Tiffany & Co.

[8] Situation Management Systems v. ASP Consulting LLC, 90 USPQ 2d 1095 (1st Cir. 2009).

expression, they are protected by copyright law. However, as with all works protected by copyright, the underlying methods and procedures are not protected by copyright.

Software owners often want to ensure secrecy and copyright allows such secrecy. When an owner needs secrecy of works such as software source code, it can register the work by providing "sufficient identifying material," instead of the normally required copy of the work. The identifying material may have confidential portions redacted to preserve secrecy.

4.2 Ownership/Authorship

Three simple rules control the initial ownership of copyright. First, in general, ownership of copyright "vests initially in the author or authors of the work."[9] Thus, a person who creates an original work automatically becomes its copyright owner. Second, and a major exception to the first rule, where a work is made for hire, the employer or commissioning party is deemed to be the author. Third, where there is more than one author, "the authors of a joint work are co-owner of copyright in the work."[10]

Although these rules are relatively simple, determining authorship can be difficult. When authors cannot agree on authorship, the Copyright Office will register conflicting claims to copyright in the same work. To avoid such a dispute, it is always best to try to resolve authorship/ownership issues at an early stage. For an employer, employment contracts should be used to define scope of employment and any work being done by independent contractors should be covered by an agreement that assigns any copyrights.

4.2.1 Works for Hire

The Copyright Act in 17 U.S.C. § 101 defines:
   A "work made for hire" is—

(1) a work prepared by an employee within the scope of his or her employment; or
(2) a work specially ordered or commissioned for use as... [specific list of types of items]... if the parties expressly agree in a written instrument signed by them that the work shall be considered a work made for hire.

Employers need not have an agreement with their employees to create a "work made for hire" obligation. Employment contracts, however, are often useful in

---

[9] 17 U.S.C. § 201(a).

[10] 17 U.S.C. § 201(a).

settling questions involving whether a particular work was prepared within the "scope of employment." Clear policies often avoid problems in this respect. On the other hand, the law requires specially ordered and commissioned works to be identified in writing as "work for hire." Additionally, even if there is a written agreement as to "work for hire" status, only the specifically enumerated types of works are eligible to be considered "work for hire."

4.2.2 Jointly Authored Works

17 U.S.C. § 101 of the Copyright Act defines a "joint work" as "a work prepared by two or more authors with the intention that their contributions be merged into inseparable or interdependent parts of a unitary whole." The authors of a joint work may be natural persons, persons, or other entities by virtue of "works for hire" or a combination of both where some contributions are "work for hire" and other contributions are not "works for hire." In the case of multiple employees creating a work for the same employer, there is only one author, the employer, so that such a work is not a joint work.

For works that are jointly owned, unless there is an agreement to the contrary, any joint owner is free to exploit the copyright in the entire work. The other joint owners, however, have a right of contribution to the profits made from the exploitation of a work.

4.2.3 Copyright Transfers: Assignment and Licensing

After initial ownership of copyright is established through authorship, copyrights may be transferred. The law requires a writing, however, to assign a copyright. 17 U.S.C. § 204(a) provides:

> A transfer of copyright ownership, other than by operation of law, is not valid unless an instrument of conveyance, or a note or memorandum of the transfer, is in writing and signed by the owner of the rights conveyed or such owner's duly authorized agent.

Proper recordation of copyright assignments with the Copyright Office should be made to perfect title to copyrights. The requirement for a writing under 17 U.S.C. § 204 does not extend to copyright licenses. Accordingly, there can be oral and implied copyright licenses. Recordation of license rights is not required. In the case of works not made for hire, both assignments and licenses can be terminated after 35 years.[11]

---

[11] 17 U.S.C. § 203.

4.3 Derivative Works

One of the informational requirements in registering copyright in a work is to identify preexisting copyrighted material that is present in the copyright application. This is important for registration purposes because copyright in a derivative work is limited to the newly-added material; the preexisting material is protected by its own prior copyright. To the extent a work "unlawfully" uses preexisting material, the law may invalidate copyright protection.

Example: Fanfiction is a popular form of writing where fanatics of a popular TV series, book or anthology of films write original stories incorporating elements from the source material such as characters, settings or situations from the TV series, book or films. Fans usually will take their favorite characters from the original source material and engage those characters in alternate or new adventures. Fanficition is legally a derivative work, and distributing fanfiction to others, for example through a fan website, is a violation of the original author's copyrights. While fanfiction is often intended as tribute, it is illegal without permission from the original content's author.

4.4 Fair Use

The Copyright Act permits certain copying under the doctrine of fair use. The Copyright Act defines a "fair use" balancing test in 17 U.S.C. § 109 that provides:

...the fair use of a copyrighted work, including such use by reproduction in copies or phonorecords or by any other means specified by that section, for purposes such as criticism, comment, news reporting, teaching (including multiple copies for classroom use), scholarship, or research, is not an infringement of copyright. In determining whether the use made of a work in any particular case is a fair use the factors to be considered shall include—

(1) the purpose and character of the use, including whether such use is of a commercial nature or is for nonprofit educational purposes;
(2) the nature of the copyrighted work;
(3) the amount and substantiality of the portion used in relation to the copyrighted work as a whole; and
(4) the effect of the use upon the potential market for or value of the copyrighted work.

The fact that a work is unpublished shall not itself bar a finding of fair use if such finding is made upon consideration of all the above factors.

What does and does not constitute "fair use" is often subject to debate, and, therefore, a lot of litigation. A good rule of thumb is that if you would object to someone copying what you are considering, you may not want to copy without consulting an attorney. A few examples of what may constitute fair use include:

- small excerpts in a review or criticism for purposes of illustration or comment;
- a parody which incorporates some elements of the work being parodied;
- quotations from a speech in a news report; and
- limited copying for use by a student for educational purposes.

Where significant use of a third party's work is desired to be made, it is generally advisable to obtain permission or a license. For music, the Copyright Act provides for compulsory licensing. For major projects, clearance activity may entail contacting multiple parties from collective rights groups such as the Copyright Clearance Center, ASCAP, and BMI to track down individual authors.

4.5 Registration Issues

Copyright registrations provide an invaluable source of information about copyrighted works and aid in the preservations of works and enhancement of the Library of Congress's collection. Moreover, for copyright owners, the registration process provides a valuable procedure to compel an examination of the issues of copyright ownership and the use of third party materials. Copyright registration is a deceptively simple procedure that leads to questions of authorship and ownership. Registration of copyright preserves important remedies, including statutory (mandatory) damages and attorney fees; in fact, a plaintiff cannot even bring a copyright action without a copyright registration.

Chapter 5
Domain Names

5.1 Securing Intellectual Property in Domain Names

Domain names are essential to promoting corporate identity and product awareness in the modern era and should be regarded like any other valuable corporate assets. A domain name is a string of unique characters used as an address to identify a particular computer or server on the Internet. For example, *epa.gov* is used to identify the United States Environmental Protection Agency website, *uspto.gov* is used to identify the United States Patent and Trademark Office website, and *siemens.com* is used to identify the Siemens Company global website.

Domain names consist of a number of domain levels. For example, in a two-level domain name such as *epa.gov*, the portion of the domain name to the right of the period (i.e., "gov") is the top level domain (TLD), and the portion of the domain name to the left of the period (i.e., "epa") is the second level domain (SLD). Many Internet users also recognize the three letter string "www." preceding the domain name. This portion of the domain is typically considered a "subdomain" which is selected by the host computer. The second level domain is usually selected by the user, and is typically used as a source identifier in the domain name. The second level domain can consist of a trademark or service mark. The most common top level domains are termed global or generic TLDs (gTLDs). The most recognized gTLD's are identified below:

- *.com*—commercial enterprises
- *.net*—networks
- *.org*—non-profit organizations
- *.biz*—businesses
- *.edu*—educational institutions
- *.gov*—U.S. government entities
- *.mil*—U.S. military
- *.int*—international organizations

In order to secure a domain name, it must first be registered with an ICANN accredited registrar. The Internet Corporation for Assigned Names and Numbers

G. B. Halt, Jr. et al., *Intellectual Property in Consumer Electronics,*
Software and Technology Startups, DOI: 10.1007/978-1-4614-7912-3_5,
© Springer Science+Business Media New York 2014

(ICANN) is responsible for the global coordination of the Internet's system of unique identifiers, including domain names. There are over 150 registrars accredited by ICANN which can be found at *icann.org/registrars/accredited-list.html*. Accredited registrars and domain name holders must implement and follow ICANN's Uniform Domain Name Dispute Resolution Policy (UDRP). Domain names are typically granted by registrars on a first come, first served basis. Registrars are not responsible for determining whether a domain name registrant has the right to obtain a domain name. For example, if the registered domain name incorporates a trademark owned by another, the domain name registrant could be liable for trademark infringement. Therefore, it is equally important that registrants conduct a clearance search prior to adopting or using a domain name.

Registering a domain name with a registrar does not grant trademark protection for the domain. In order to seek trademark protection for a domain name, the owner must file an application with the U.S. Patent and Trademark Office. In order to obtain a federal trademark registration for a domain name, the applicant must satisfy all of the legal requirements for registrable trademarks. In addition, the U.S. Patent and Trademark Office has specific procedures for applications to register domain names. In registering a domain name as a trademark, the top level domain (e.g., .gov, .com, etc.) and subdomain (e.g., www.) are usually ignored. The USPTO typically considers only the second level domain when examining the mark for likelihood of confusion with other marks. Where the domain name is used only as an Internet address, and is not used to identify the source of goods or services, the trademark is not registrable. However, if the SLD is an actual tradename for a product, or the domain name owner uses the domain name to advertise its goods or services, the U.S. Patent and Trademark Office is more likely to find the trademark registrable.

Example: Using a company's trademark as a domain name can be a very useful business tool. For example, the Sony Corporation uses its registered trademark Sony® as a domain name to help customers identify and easily remember its web address www.sony.com. Sony already owns the trademark rights to the Sony® mark and the domain name identifies the source of the goods which it sells on its site. Conversely, a domain name that incorporates generic terms cannot seek trademark protection, because the generic terms would not be registerable as a trademark on their own.

Domain names used to advertise a good or service can obtain trademark protection. Hypothetically, if Sony launched a new PlayStation videogame console, the PlayStation 5, Sony could register the domain name www.playstation5.com and it is likely that the domain name would be considered registrable by the USPTO as a trademark, so long as the mark satisfied the remaining requirements for trademark applications since it uses the actual name for Sony's underlying new videogame console product.

5.2   Domain Name Disputes

Domain name disputes can arise in a number of ways. For example, use of another's trademark in a domain name can be subject to an action for trademark infringement, unfair competition, or dilution of a trademark. "Cybersquatting" is another type of act that has been subject to dispute. Cybersquatting is typically considered the act of registering a domain for the purpose of preventing a trademark owner from using it in order to extract payment from the trademark owner. In addition to holding the domain name hostage to the trademark owner, cyber-squatting also includes situations in which domain name registrants have registered domain names that incorporate trademarks with the intent to benefit (e.g., by way of advertisements) from inadvertent traffic at the registrant's website.

In order to address cybersquatting, Congress enacted the Anticybersquatting Consumer Protection Act (ACPA). The ACPA identifies cybersquatting as:

(A)   A person shall be liable in a civil action by the owner of a mark, including a personal name which is protected as a mark under this section, if, without regard to the goods or services of the parties, that person—

   (i)   has a bad faith intent to profit from that mark, including a personal name which is protected as a mark under this section; and
   (ii) registers, traffics in, or uses a domain name that—

   (I)   in the case of a mark that is distinctive at the time of registration of the domain name, is identical or confusingly similar to that mark;
   (II) in the case of a famous mark that is famous at the time of registration of the domain name, is identical or confusingly similar to or dilutive of that mark; or
   (III) is a trademark, word, or name protected by reason of section 706 of title 18, United States Code, or section 220506 of title 36, United States Code.[1]

Thus, in order for a trademark owner to bring a claim under the ACPA, the owner must establish: (1) that the mark is distinctive or famous; (2) the domain name registrant acted in bad faith by use of the mark; and (3) the domain name and the trademark are either identical or confusingly similar.

Around the same time that the ACPA was enacted in the United States, ICANN developed the Uniform Domain Name Dispute Resolution Policy (UDRP) to similarly address domain name abuse that impacts trademarks owners world wide. The goal of the UDRP is to create a lower cost administrative process for the resolution of domain name disputes. In a UDRP proceeding, the trademark owner must prove the following three elements:

(1)   that the domain name at issue is identical or confusingly similar to a trademark in which the complainant has rights;

---

[1]   15 U.S.C. § 1125(d)(1)(a).

(2) that the domain name registrant has no rights to, or a legitimate interest with respect to, the domain name; and

(3) that the domain name has been registered and is being used in bad faith.

While a UDRP proceeding is beneficial in that it can result in a speedy and lower cost disposition of a domain dispute, the only available remedies in a UDRP proceeding are the cancellation or transfer of the disputed domain name. As discussed above, whenever a registrants signs an agreement for registration of a generic or global TLD (e.g., .com, .net, .org, etc.), the registrant must agree to resolve any disputes with third parties regarding the domain name under the UDRP process.

5.3 Case Study: Virtual Works, Inc. v. Volkswagen of America, Inc.: The Dispute of vw.net

In 1996, Virtual Works registered the domain name www.vw.net with Network Solutions, Inc., which at the time was the only company authorized to serve as a registrar for Internet domain names.[2] Virtual Works' two proprietors, Christopher Grimes and James Anderson, were aware at the time of registering the domain name that customers might think that the site was associated with Volkswagen due to the famous nature of Volkswagen's "VW" trademark.[3] The two even discussed how Volkswagen might one day approach Virtual Works to work out a deal to buy the domain name from them.[4]

In 1998, several Volkswagen dealers approached Virtual Works about buying the domain name.[5] With the increased number of persons interested in buying the domain name, Virtual Works approached Volkswagen regarding a sale of the domain name.[6] In an unorthodox fashion, Anderson reached out to Volkswagen's trademarks department by leaving a voice message indicating that he owned the rights to vw.net and that unless Volkswagen agreed to purchase the domain name, Virtual Works would sell the domain to the highest bidder.[7] Volkswagen responded by initiating dispute resolution procedures with Network Solutions.[8] Network Solutions informed Virtual Works that the domain name would be lost if Virtual Works did not seek declaratory judgment against Volkswagen;[9] (declaratory judgment being a legal determination made by a court as to the legal position of litigants in the case when there is doubt as to their position in the law).

---

[2] Virtual Works, Inc. v. Volkswagen of America, Inc., 238 F.3d 264, 266 (4th Cir. 2001).

[3] Id.

[4] Id.

[5] Id. at 267.

[6] Id.

[7] Id.

[8] Id.

[9] Id.

Both the District Court, and the United States Court of Appeals for the Fourth Circuit held that Volkswagen's challenge alleging that Virtual Works' use of the domain name www.vw.net was an attempt by Virtual Works to one day sell the domain name to Volkswagen was a violation of the ACPA.[10] Virtual Works was cybersquatting on Volkswagen's trademark rights by using Volkswagen's trademark as part of the Virtual Works domain name. The Fourth Circuit Court of Appeals even went as far as to create a common law requirement that the cybersquatter must exhibit a bad faith intent in order to confer liability. Domain names that bear a close resemblance to a trademarked name are not *per se* impermissible, but instead the domain name must have been registered with the bad faith intent to later sell the domain name to the trademark holder.

Volkswagen then sought use of the www.vw.net domain name for itself, which is the remedy called for by the ACPA when the act is violated by a cybersquatter.[11] The court may order "the transfer of the domain name to the owner of the mark."[12] The ACPA was enacted to prevent the expropriation of protected marks in cyberspace and to abate consumer confusion resulting therefrom.

5.4 New Generic Top Level Domain Names in 2013

Since the inception of the Internet new generic Top-Level Domain Names (gTLDs) have been introduced for use. At the close of May 2012, ICANN concluded the application phase of an initiative designed to expand the Internet's Domain Name System. The initiative is called the "New gTLD Program" and ICANN has accepted thousands of applications, submitted by companies, for new gTLDs for Internet addresses (gTLDs being the .com, .gov, .edu, suffix attached to the end of a domain name).

Hundreds of new gTLDs have been introduced, including .baby, .mail, .search, .tickets, and .wedding. Internet-based companies, such as Google or Amazon, applied for 101 gTLDs and 76 gTLDs respectively. However, the popular social media sites such as Facebook and Twitter did not participate in the application process. Some companies purchased their own name as a gTLD, including Toshiba, Samsung, Microsoft, Panasonic and Verisign.

New internet addresses could look something like this:

- www.amazon.shop
- www.jnj.baby
- www.aol.patch
- www.gap.piperlime

---

[10] Id. at 270–271.

[11] Id. at 271.

[12] 15 U.S.C. § 1125(d)(2)(D)(i).

A complete list of the applied-for gTLDs and the company applicants that applied for them can be found at http://newgtlds.icann.org/en/program-status/ application-results/strings-1200utc-13jun12-en.

The process for obtaining a new gTLD was limited to a 3 month window of opportunity to apply spanning from January 12, 2012 to April 12, 2012. The process included filing an application for the choice gTLD and paying an application fee of $185,000 fee per submission. As a result of the application process, duplicates of certain gTLDs were inevitable, i.e., five companies applied for the gTLD.style and four companies applied for the gTLD.radio. The application process also poses the potential problem of proper ownership of the gTLD. It is possible that an applied-for gTLD is identical to, or confusingly similar to, an existing trademark owned by another. Granting the gTLD to someone other than the trademark owner would infringe the legal rights of the trademark owner.

To resolve these issues, ICANN initiated a public review process for all of the applied-for gTLDs. Applications are evaluated in batches. Once the application is selected for evaluation, the public has a 60-day window to make any comments and a 7-month window to make any formal objections to the applied-for gTLD. Formal objections are limited to:

- "String" confusion objections, which involves the issue of duplicate applied-for gTLDs or those applied-for gTLDs that are confusingly similar to another applied-for gTLD;
- Legal rights objections, i.e., a trademark owner (other than the applicant) has legal rights to the mark being used as a gTLD;
- Public interest objections; and
- Objections based on the interests of particular communities.

After the ownership disputes of the new gTLDs have been settled, ICANN plans to release the new domains in batches. Those companies who are granted the new gTLD names will have to pay an annual $25,000 maintenance fee to keep the rights to the domain names.

Chapter 6
Intellectual Property Issues in Labeling and Marketing

Apart from whether a mark can be protected under the trademark law, there may be government regulations that can restrict or prohibit use of the mark in advertising or labeling any product(s) with the mark. Anyone using or selecting a trademark in the high-tech consumer electronics or computer software industries should be aware of the nature and kind of regulations that may be applicable.

6.1 Governmental Controls Over Advertisements and Labeling

The production, marketing, distribution, import, export and sale of high-tech consumer electronics and computer software products in the United States are subject to government regulations on both the federal and state levels.

6.1.1 Federal Trade Commission

The Federal Trade Commission ("FTC") is responsible for maintaining a competitive marketplace for both consumers and businesses and preventing unfair or deceptive trade practices. As such, it administers laws and regulations ranging from the content of clothing labels to laws requiring truth in advertising and prohibiting price fixing.

The FTC regulates advertising under its statutory authority to prohibit deceptive acts or practices. The FTC will find an advertisement deceptive (1) if it contains a representation or omission of fact, (2) that is likely to mislead consumers acting reasonably under the circumstances, and (3) that representation or omission is material.

The first step in the analysis is to identify representations made by an advertisement. A representation may be made expressly or implicitly. An express claim

G. B. Halt, Jr. et al., *Intellectual Property in Consumer Electronics, Software and Technology Startups*, DOI: 10.1007/978-1-4614-7912-3_6, © Springer Science+Business Media New York 2014

directly makes a representation of fact. An implied claim is not so straightforward and requires an examination of both the representation in order to determine the overall meaning or commercial impression of an advertisement. False claims can also stem from an omission of information, which makes an affirmative representation misleading. In other words, it can be deceptive for a seller to simply remain silent if such silence constitutes an implied, but false, representation.

The second step in identifying deception in an advertisement requires the Commission to consider the representation from the perspective of a consumer acting reasonably under the circumstances.

Finally, a representation must be material, i.e., likely to affect a consumer's choice or use of a product or service. Express claims involving health, safety, price or efficacy are presumed material.

Any claim by an advertiser must have a reasonable basis. Where compliance claims are made (i.e., Energy Star compliance or European Conformity markings), those claims should normally be substantiated by competent and reliable scientific evidence such as tests, analyses, research, studies or other evidence conducted and evaluated in an objective manner by persons qualified to do so, using generally accepted scientific methods.

The FTC also monitors and reports on industry practices regarding the marketing of violent movies, music, and electronic games to children; provides guidelines regarding US origin claims placed on product packaging; and guidelines on the use of environmental marketing claims, or the marketing of "green" products.

6.1.2 State Regulation

Individual states within the U.S. also regulate food labeling and advertisements. Many activities, but by no means all, relating to labeling are preempted by federal laws and regulations promulgated by the FTC and other agencies. The strength of state power lies in each state's consumer protection laws which seek to prevent deceptive trade practices. Since the penalties for violating these laws can be severe in some cases, individual states can have a great influence on regulating advertising of products.

States can also regulate how consumers dispose of their electronics. For example, some states have enacted laws that prohibit disposal of consumer electronics in state landfills, and have instituted state-regulated electronics recycling programs. Of note are laws such as "Producer Take Back laws," which require manufacturers of certain electronic devices to take back particular types of old electronics that they were responsible for selling in the first place. Companies should take care to ensure that they do not run afoul of the laws.

6.1.3 Governmental Controls Outside of the United States

Regardless of one's knowledge of U.S. labeling laws and regulations, these laws and regulations are only valid within the jurisdiction of the United States. If a consumer electronic device or computer software is intended to be exported, the labeling requirements of the import country must be identified and followed. Furthermore, the U.S. Bureau of Industry and Security has policies and regulations for the export of dual-use commodities (dual-use items are predominantly for commercial uses, but also have military application), software, and technology.

By way of example, the European Union (EU) requires the European Conformity (CE) marking be affixed to any product subject to it before the product is to be placed on the market in the EU. Similarly, the EU, China and Japan each require markings on products that consist of certain hazardous substances (Restriction on certain Hazardous Substances or "RoHS") to indicate compliance with each countries' RoHS standards.

6.1.4 Patent Marking

The America Invents Act (AIA) has changed the manner in which products may be marked with patent numbers. The purpose of placing patent numbers on a product is to put purchasers on notice as to what patents protect the product. Marking serves as "constructive notice" to potential infringers. The AIA amended 35 USC § 287(a) to allow products to be "virtually marked." Marking with "patent" or "pat." and "an address of a posting on the Internet, accessible to the public without charge for accessing the address, that associates the patented article with the number of the patent," is now a permissible form of notice.

> *Example*: Under the AIA, a company's product could be virtually marked in any of the following ways:
>
> Patents: http://www.companyname.com/patents
> Patents: http://www.companyname.com/patentnotice
> Patents: http://www.companyname.com/legal/patent

Prior to the adoption of the AIA, an individual could bring a suit on behalf of the U.S. in what is called a qui tam action, for "false patent marking," where products were marked with expired patent numbers. The AIA introduces 35 USC § 292(c), which effectively eliminates any causes of action based on marking products with expired patent numbers.

6.2 Non-Governmental Controls

Many industries have trade associations that may have a code of conduct governing advertisements of their products. However, any such code of conduct is effectively non-binding on members. Moreover, if the code of conduct limits information provided to consumers (e.g., in the event of truthful competitive advertising) it could be considered anti-competitive and subject to challenge by the FTC. Examples of this would be an industry code of conduct that imposes a higher standard of substantiation for comparative claims than for unilateral claims.

Alternatively, false or misleading statements can be brought before the National Advertising Division ("NAD") of the Better Business Bureau (www.nadreview. org). In an action before the NAD, a competitor may file a complaint to take action against false or misleading statements, while avoiding the distractions and expense of full blown inter-party litigation or involvement with a government agency. After a complaint is initiated and accepted, the NAD will investigate the matter, and if a deceptive or misleading practice is found, it will take action which can range from requesting corrective advertising from the publisher, publication of the NAD's findings, or referral to a regulatory authority such as the FTC for further penalties.

6.3 Comparative Advertising

Comparative advertising refers to identifying a competitor's product by its trademark and comparing it to the advertised product. Comparative advertising is widely used in the high-tech consumer electronics and computer software industries to convey valuable information to consumers. However, using such comparative advertising runs the risk that a competitor could take action that the claims are either false or that the advertisement infringes the competitor's trademark rights.

The FTC and courts have approved the use of brand comparisons where the bases of the comparison are truthful, objective and clearly identified. Advertisements containing truthful and non-deceptive statements that a product has certain desirable properties or qualities which a competing product does not possess are permitted. However, false or misleading comparative advertising can trigger liability for false advertising or trademark dilution. Therefore, when a competitor's trademark is used in comparative advertising, the statements used must not have a tendency or capacity to confuse or be considered to be false or deceptive.

In order to lessen the risk of litigation arising from comparative advertisement, the steps outlined below should be considered before making any claims regarding a competitor's product.

- Only use the competitor's trademark to the extent necessary for comparison purposes.
- Use the competitor's trademark in the same way that the text is used throughout the advertisement. In other words, do not place emphasis on the competitor's trademark that would lead a consumer to believe that the trademark is associated with the advertiser's company.
- Do not disparage or mock the competitor or the competitor's trademark in the advertisement. Make sure any photographs of a competitor's product are accurate and do not place the product in unfavorable light. In addition, do not use any photographs of a competitor's product that may be copyrighted by a competitor or a third party. Instead, use original photographs.
- Try to make sure your product's advertisements, packaging and trade dress create a separate commercial impression from that of your competitor. In other words, a consumer looking at the two products or advertisements should not believe that the products are somehow related because of similarities in colors, fonts, stylization, etc.
- Include an easily visible disclaimer in any advertisement that uses the competitor's trademark. An example of such disclaimer is, "COMPETITOR'S TRADEMARK is a trademark of COMPETITOR. ADVERTISER does not make or license COMPETITOR'S TRADEMARK and is not affiliated or associated with Competitor."
- Verify that any statements regarding competitor or advertiser's product are true both explicitly and implicitly. The following steps should be taken:
  - Retain a certified, independent and well-respected laboratory in the industry to perform the testing. Verify that the testing to be performed is the test generally accepted by technical or scientific community.
  - Make sure testing is fair and impartial. For example, ensure that the appropriate comparative products from the competitor's product line are compared.
  - All statements regarding a competitor's product should be supported by testing.
  - Make sure copies of the test methodology, test results and samples tested are preserved in a safe place in case the tests have to be repeated at a later date.

Even if all the above suggestions are followed, there is nothing that can be done to prevent a party that feels it has been harmed from deciding to bring a court action to protect its rights. It is recommended to consult legal counsel regarding individual advertising claims before a comparative advertising campaign is launched.

6.4 Case Studies: Comparative Advertising: Smartphones and Service Providers

6.4.1 Battle Between Smartphones: Apple's iPhone Versus Samsung's Galaxy S II

Two major competitors in the smartphone industry are Apple, with its iPhone[®1] product, and Samsung, with its Galaxy S[®2] product. The phones look remarkably similar, and perform similar functions in similar ways. They each have a full touch screen, application icons that look similar to each other, and even have similar shape—a rectangular-shaped phone with rounded corners. Marketing such a similar product to the iPhone, Samsung has adopted a comparative advertising campaign, designed to steal market share from competitors, primarily Apple. The advertising campaign is directed at a target market of young people who consider themselves hip, cool and creative. Coincidentally, this is also the iPhone's target market.

Both the Samsung Galaxy S II and Apple's iPhone 4 were launched in February of 2011. When Apple released the next upgrade—the iPhone 4S—in November of 2011, Samsung pursued a viral comparative advertising campaign designed to draw attention to the iPhone's shortcomings, and highlight the Galaxy S II's superior functionality. The ad campaign was called the "Next Big Thing" with the tag line "the next big thing is here" implying that Samsung's Galaxy S II is the most advanced smartphone available. The ad shows long lines of young people waiting to pick up their new smartphone. The people are commenting about how long they have waited and how excited they are to get their hands on the new smartphones. This visual of people standing in line brings to mind images from news reports about Apple loyalists waiting in line for days to get their new iPhones. In the ad, a Samsung smartphone owner enters the scene, using their Galaxy S II and the people waiting in line are captivated by it. In the ad, the Galaxy has a large screen, is 4G enabled and has an assortment of other desirable features. The ad features a visual side-by-side comparison of the products; showing the Galaxy S II and an iPhone next to each other. This is the only evidence of an iPhone in the entire ad. In the entirety of the ad, it is never explicitly stated that the Galaxy S II is being compared to an Apple iPhone. Samsung is careful to never use the Apple logo or Apple's name in the ad at all. Samsung doesn't even use the word "iPhone" in any part of the ad. Regardless, it is apparent from all the visual clues in the ad that an iPhone is the competitor's product.

Samsung has been careful in its comparative advertising campaigns. It has refrained from using Apple trademarks, such as its logo or brand name, in its advertisements. Samsung has also been careful to only draw comparisons between the two competing products for aspects of the device which are verifiable. For

---

[1] U.S. Registration No. 3,746,840. Owner: Apple, Inc.

[2] U.S. Registration No. 3,905,843. Owner: Samsung.

example, in a subsequent advertising campaign, each of the technical specifications of the iPhone 5 and the Galaxy S III that are comparable is a true statement, which can be tested by an independent third party for verification. The screen on the Galaxy S III is larger than the screen on the iPhone 5, the Galaxy has more RAM, the Galaxy S III has longer battery life; all these statements are verifiable as true, and as such, can be used by Samsung for comparative advertising purposes.

6.4.2 Same Product, Different Service Providers: AT&T's iPhone Versus Verizon's iPhone: A Battle for Market Share

It is common to see comparative advertising between two different products, for example Samsung's Galaxy S smartphone compared to Apple's iPhone discussed in the previous case study. However, service providers can also engage in comparative advertising in an attempt to secure market share over users of a specific product. This case study examines the competition between AT&T and Verizon to provide cellular service to iPhone users.

For many years, AT&T and Verizon have been embroiled in a comparative advertising battle over which company had the better network coverage. When it comes to 3G coverage, the public perception, whether warranted or not, was that Verizon coverage was better than AT&T's coverage. However, Verizon was missing out on a major segment of business that only AT&T had access to—the iPhone customer base. Since 2007, Apple and AT&T had an exclusive contract where all iPhone purchasers were required to subscribe to AT&T's wireless service. Then in 2011, Verizon announced that it would also carry the iPhone. Availability of the iPhone from Verizon spurred a flurry of new comparative advertising between AT&T and Verizon, battling over iPhone market share and AT&T defectors. A few points of differentiation between the service providers regarding the iPhone include multifunction capabilities of the phone and which provider had the better network.

Verizon's 3G network operates on a network standard called CDMA2000. The deployed version of Verizon's CDMA2000 network does not allow for the simultaneous use of voice communication and data transmission. AT&T quickly latched onto this deficiency in Verizon's network capability by running advertisements showing AT&T iPhone customers surfing the Internet while on a phone call. Verizon countered by returning to what it knows it does best, and ran advertisements emphasizing Verizon's superior network coverage and better reliability over AT&T. Taking a modification on its traditional Verizon network test man tagline "Can you hear me now? Good," the Verizon iPhone network test commercial has the test man holding an iPhone to his ear, saying "Yes, I can hear you now." The implication being that Verizon iPhone users experience better coverage than AT&T iPhone users.

Part II
Implementing Intellectual Property Practices, Procedures and Strategies

Chapter 7
Seven Basic Steps to Getting Started

7.1 Confidential Disclosure or Non-Disclosure Agreements

Confidential Disclosure Agreements, which are also referred to as Non-Disclosure Agreements or NDAs, refer to a contract that protects confidential or trade secret information ("Confidential Information") from disclosure to third parties. NDAs are commonly included as a part of an employment contract, and are also included in, or form, a separate agreement with vendors, contractors, and sometimes customers.

An NDA should include the following provisions:

a. A clear definition of the information that is to be held in confidence, as well as a provision defining how the information should be marked or identified, for example with a "Confidential" stamp or label. Employment agreements typically adopt broader definitions requiring employees to maintain any work-related information that the employee develops or has access to as confidential. Supplemental agreements for specific employee development projects may also be used to more clearly identify the information that is subject to the agreement.
b. A specific recitation of the limited purposes for which the Confidential Information can be used. For example, in an NDA with an outside vendor, the Confidential Information may be provided for specific testing and evaluation of a product, such as RF transmitter and receiver calibration testing. For an advertising or marketing agency, the Confidential Information could be limited to the specific purpose of package development, branding, or other specific marketing tasks for the benefit of the disclosing party.
c. A recitation that the receiving party cannot breach the confidential relationship, induce others to breach it, or induce others to acquire the Confidential Information by improper means.
d. A recitation of exceptions to the confidentiality requirements. This generally will exclude any information that: (i) the receiving party can show they were already in possession of at the time of the disclosure; (ii) information that is in the public domain through no fault of the receiving party; and (iii) information that the receiving party receives without restriction from a third party.

G. B. Halt, Jr. et al., *Intellectual Property in Consumer Electronics,*
Software and Technology Startups, DOI: 10.1007/978-1-4614-7912-3_7,
© Springer Science+Business Media New York 2014

e. The time period that the information must be held in confidence. This can be any reasonable term agreed upon by the parties, and often falls in the range of 2–5 years, depending on the technology in question and the disclosure's purpose. For example, an advertising agency that is working on an advertising campaign that is going to be released within a year would not need an NDA term that extends beyond the advertising campaign release. However, for an outside consultant that does product testing of beta products to determine a preferred version of a product for commercialization, the NDA could justifiably require that the information be held in secrecy for 5 years or more.

f. A provision defining the remedy available to the non-breaching party in the event of a breach or impending breach. This should recite that the disclosing party is entitled to injunctive relief for breach of contract by the receiving party to prevent the release of the Confidential Information. It is also possible to include a liquidated damages provision as an incentive for the receiving party not to disclose the Confidential Information, although this is less typical given the circumstances surrounding most NDAs where the disclosing party is attempting to obtain information or services from the receiving party.

g. Other miscellaneous provisions can include: (i) a provision for return or destruction of the Confidential Information when the task or review by the receiving party is completed; (ii) a transfer of ownership of any additional intellectual property that results directly from the Confidential Information and/or in connection with the work being done by the receiving party; and (iii) a recitation of the courts or jurisdictions where any potential dispute will be resolved.

While NDAs have several advantages, there are a few drawbacks that they cannot address. Many large companies will not sign NDAs for any outside submissions, regardless of purpose. In fact, some large corporations require the opposite: a signed statement saying that no information will be held in confidence and that the party submitting the information will rely exclusively on any intellectual property rights (such as patent or trademark rights) that they applied for or may obtain as the sole recourse against the receiving party in the event that their information is used. This typically occurs as a result of a competitor's parallel development efforts. A competitor may accuse a company receiving its confidential information of theft or misappropriation even though the receiving company was already working on a similar development. Rather than face potential law suits or bad publicity, the policy of such express waivers of confidentiality shields the receiving company.

Even if an NDA is signed, if the Confidential Information is purposefully or even inadvertently disclosed, it may not be possible to "put the genie back in the bottle." Although damages may be available against the discloser, an actual public disclosure cannot be undone. This could result in inadvertent loss of the ability to seek patent protection in many countries that have absolute novelty requirements. While the United States allows a one year grace period from the date of first public use or disclosure of an invention, by the inventor, to the public, if the owner of the Confidential Information is unaware of the disclosure, U.S. Patent rights could also be lost. Furthermore, any trade secret information that is publicly disclosed ceases to be a trade secret.

Example: A company should execute employment agreements with all employees who will have access to a new project, including its senior engineers, designers, programmers, etc. as well as anyone else that will be working on the new product, its marketing employees, and any others who will have access to the information. If confidentiality terms are not included in the employment agreements, the company can enter separate agreements, preferably at the same time as an annual review and pay raise so that there is no question regarding a potential lack of consideration for the new NDA provisions. The confidentiality term should extend for a time period beyond the end of employment, especially if a rival competitor is known to poach employees away from the company in order to copy products or product concepts. In the event that an employee with knowledge and/or confidential information related to the project leaves the company, the NDA should provide for injunctive relief to prevent the confidential information from being improperly disclosed.

In addition, any outside vendors used for the new project should sign a project specific NDA with the company. The vendor agreements should also have specific terms spelling out ownership of any new developments made using the confidential information while carrying out the work for the company.[11]

A distribution of a new product will likely require advance notice of the product. If the company learns that its distributor will not sign an NDA, and to the contrary requires a specific waiver of confidentiality to accompany any offer or disclosure, the company should have any patent applications for its new product filed with the U.S. Patent and Trademark Office before any disclosure of any information to the distributor. This will protect any potential U.S. or foreign patent rights that the company may choose to pursue.

7.2   Assignment of Rights

An Assignment is a contract between two parties in which rights owned by one party are assigned to the other party. In the U.S., all rights to an invention are initially vested in the inventor(s). Accordingly, a formal Assignment is important to transfer rights from an inventor or inventors to the company. For employees, the obligation to assign inventions made in the course of an employee's regular job duties can also be included in the employment contract. If the obligation to assign inventions is not included in an employee's employment agreement, a separate

---

[1]   A sample NDA is set forth in Appendix A.

agreement can be entered including the obligation to assign inventions to the company. This separate agreement should preferably be made and signed with the employee in connection with an annual review and raise so that there is no issue with respect to consideration for the agreement.[2]

Many companies offer employee incentive programs that pay bonuses for making inventions that help the company. This typically takes the form of a lump sum payment at the time a patent application is filed or a patent is granted. Some countries, such as Germany, have specific statutory requirements that define the amount that an employee must be compensated for any invention that is used by the company.

When working with third parties, it is common to include assignment terms in the vendor contract for inventions related to the specific work. For patentable inventions, this can be critical if a breakthrough is made by the vendor in connection with the development of a company's product. If no agreement is reached before the contract work is undertaken, rights would belong to the vendor or vendor's employee, creating the potential for being forced into a sole source of supply, or worse, having the product or flavor which the company paid the vendor to develop offered to third parties without any ability for the company to control it.

Example: Assignments are critical for any intellectual property. For example, assume a technology company hires a contractor (writer, drafter, programmer, designer, etc.) whose work product would be eligible for copyright protection. In the absence of a written assignment for works made under contract, the copyrights would be owned by the contractor. It must be clear from the hiring contract that any IP rights created by the contractor are assigned to the technology company and that any work done by the contractor was done as a "work for hire," thereby assigning all copyrights to the tech company.

7.3 Employee Education

Employee education is one of the cornerstones of a successful intellectual property program. Inadvertent disclosure of confidential information or new inventions that are being developed can easily result in any potential intellectual property rights being lost, or worse, landing directly in a competitor's hands. The only way to effectively address this is to train employees such that they understand the ramifications of their actions and the potential cost to the company in terms of lost

---

[2] A sample Assignment contract from an individual inventor to a company is set forth in Appendix B.

intellectual property rights and lost profits on new products or developments that cannot be protected.

Many companies hold intellectual property seminars that are presented by in-house or outside intellectual property counsel. These programs educate employees on the basic tenants of intellectual property law in a practical setting, and how they can handle particular situations. A typical program would include a review of what may constitute patentable subject matter given the company's technology field, the potential bars to patentability that an employee should be aware of, and a review of the company's system for documenting inventions and subsequently handling those documents. Employees that regularly deal with third party vendors should also receive special training related to risks associated with dealing with vendors.

7.4 Accurate Record Keeping

7.4.1 Patents

In the United States, patent applications filed on or after March 16, 2013 are subject to the America Invents Act "first to file" rule, meaning that the right to the grant of a patent lies with the first person to file a patent application, regardless of the date of actual invention. There is a public disclosure grace period, however, which allows the inventor to disclose the invention up to one year prior to filing. In the event that an inventor publicly discloses the invention before another inventor independently develops the same invention and files for a patent on the invention, the first inventor may be entitled to the patent protection due to the one year disclosure grace period. It is important to keep detailed records as to when a disclosure of an invention is made.

Another document that the company should have is an employee invention submission form. This should include sections for a complete description of the invention, as well as any potentially critical events that occur, such as a disclosure to others. This form serves two purposes: (1) it can serve as the vehicle for in-house review and a determination of whether patent protection is going to be pursued; and (2) if the company proceeds with a patent application, it can act as the vehicle for transmitting information on the invention to patent counsel for searching and/or the preparation of a patent application.[3]

For each invention disclosure that is ultimately pursued as a patent application, a company should open a separate file as a place to store all information related to the invention, including copies of the relevant completed pages of the inventor's notebook, the invention disclosure form, any known prior art that might relate to the invention, as well as all correspondence and documents related to the preparation, filing, and prosecution of a patent application before the USPTO.

---

[3] A sample invention disclosure form is set forth Appendix C.

Example: There are plenty of horror stories about small time inventors being duped by a large corporation into disclosing the inventor's idea and then the company steals it. A few famous examples come to mind, including the 2007 allegations that Facebook's founder Mark Zuckerberg built the Facebook social media platform from stolen code, or the patent infringement allegations against Heinz's Dip & Squeeze condiment container asserted by a Chicago man who pitched his patent-pending idea for a similar condiment container to Heinz in 2005.

7.4.2 Trademarks

Trade secret protection depends on defining and following a strict set of policies for handling the information that is being protected as a trade secret. Trade secret policies should be documented in a policy manual, and all employees that have access to the trade secret information should receive training on handling trade secret information and should be required to periodically review the company's policy regarding treatment of such information. Policies should include how to identify and mark trade secret information, including information or technology under development. Identification can be as broad as "all information related to project X" or down to a specific formula for a product, and can be set by management or counsel. Marking should be on both physical and electronic documents, using labels like "Confidential," "Secret", "Trade Secret" or "Proprietary Information" of the company.

Access to the trade secret information should be on a need to know basis inside the company. Access to the information should be on a "log in–log out" basis, whether on paper or electronically. Any electronically stored or transmitted information should be encrypted based on a defined procedure. A policy should also define storage of the information when an employee is away from his work area, and may include a locked central or private storage area. The reason for this high level of security is that in a misappropriation lawsuit, a court's inquiry will not only focus on the bad acts of the accused party, but it will also examine and consider whether the company asserting its trade secret rights adequately maintained and safeguarded its trade secrets.

Example: In terms of the number of patents issued each year in the United States, Sony has ranked as one of the top ten recipients for nearly a decade. Sony values intellectual property, and has company policies regarding procedures that employees must follow when dealing with company IP and know how. Not only does Sony require that its employees respect IP developed internally, but Sony also requires employees to respect the IP of others. All IP generated by Sony employees are assigned to the company. No employee may disclose any proprietary information unless authorized to do so by the company. Use of IP and trade secret information is limited to only those employees who have a need to know and use the protected information.

7.5 Patent and Trademark Searches

7.5.1 Patent Searches

There are a number of different patent searches that can be useful for a number of different purposes, including patentability, infringement clearance, validity, and state of the art searches.

A patentability search is the most basic search, and involves searching U.S. patents and patent publications as well as potentially other patent and non-patent literature to determine whether an "invention" meets the USPTO requirements for patentability, based on the documents identified by the search. This is a useful tool to gauge the potential for patentability; however, it is not a guarantee that a patent would ultimately be granted. The limitations on patentability searches are that they are generally not exhaustive, and other more pertinent references may ultimately be identified from areas not searched. Accordingly, while negative results can be relied upon, a positive search report merely leaves the possibility of patent protection open.

An infringement clearance search is a search of U.S. patents that remain in force for potential infringement by a new product that is being developed. Infringement clearance searches should be performed prior to the product's launch. This type of search involves a review of the patent claims that remain in force in the relevant classifications for the product being developed. As the search requires a specific review of each independent claim of the relevant patents, it is more complete. If done early enough in the design or development process, a clearance search allows the new product to be modified before potential infringement.

A validity search is a prior art search of U.S. and foreign patent and non-patent documents that is directed against the claims of a specific patent. This can be used to determine whether the claims of a known or asserted competitor patent are valid. The results of the search can be used for negotiations, or can be used to invalidate the patent in a court or USPTO proceeding.

A state of the art search looks at U.S. and possible foreign patent document collections for representative technology and developments in a particular field. This is used as a research tool for resolving a particular problem or for examining the type of work competitors have done in a given field.

7.5.2 Trademark Searches

Trademark searches determine whether a company can adopt a trademark for its product or service. A trademark search can be done in the Federal Trademark database, or can be done in one or more comprehensive databases. Trademark searches should be carried out before adopting a trademark to determine whether any third party has used the trademark for the same or similar types of goods, and may therefore have superior rights in the trademark. Adopting a third party's mark, whether knowingly or unknowingly, may result in a lawsuit for trademark infringement. A trademark search can help avoid this costly litigation.

7.6 Decide on the Type of Protection Early in the Inventive Process

The earlier a company decides what type of intellectual property to pursue, the lower the likelihood that rights will be inadvertently lost. For patents, it is important to observe specific timelines before the invention's first public disclosure, use, or offer for sale. These dates will also determine certain statutory bars to patentability in the United States. For trade secrets, the earlier that a decision is made to protect a new product, or even its method of manufacture as a trade secret, the easier it will be to ensure that some information is not inadvertently released.

7.7 Speak to an Intellectual Property Attorney

As the facts and circumstances surrounding product development and branding vary widely, consulting with an intellectual property attorney is highly recommended.

> *Example*: Virtually all major corporations who utilize and value IP rights have in-house counsel that specifically practices IP law. If a company doesn't have in-house counsel, the company certainly seeks IP counsel externally. Sony, Google, Microsoft, Red Hat, Riot Games, VMware, DISH, Comcast, Yelp, Texas Instruments, Intel and Amazon are just a handful of examples of companies that have IP counsel on staff. Even the USPTO has general IP counsel. Whether your company is just starting up, is small but growing, or is fully established and highly successful, consulting IP counsel is important for protecting your company's intellectual property rights.

Chapter 8
Deciding Between Patent or Trade Secret Protection

Many companies use both patents and trade secrets to protect inventions. However, since these two forms of protection are mutually exclusive in a fundamental regard (i.e., whether or not to disclose an invention), companies must often chose between the two forms of protection. Each has advantages over the other that should be carefully considered when forming the appropriate intellectual property strategy.

8.1 Differences in Scope of Protection

A first consideration is the difference in subject matter between trade secrets and patents. Trade secret protection covers a wider range of possible innovations and it has a potentially unlimited duration. The Uniform Trade Secret Act ("UTSA") defines trade secrets as any information, including a formula, pattern, compilation, program, device, method, technique or process that derives independent economic value from being "secret."[1] While concepts, databases, and compilations are generally not patentable, the UTSA expressly protects them if they are valuable to the business and the business takes specific steps to keep them secret. Further, other categories of information that may be subject to trade secret protection, where patent protection would not be allowable include customer lists, product pricing, strategic planning, company policies, market analyses, etc.

8.2 Differences in Litigation Remedies

A second consideration is the different types of litigation remedies. A U.S. patent grants the owner the right to exclude others from making, using, or selling the invention throughout the United States. In return for this right, the patentee must

---

[1] The UTSA is discussed above in the chapter on trade secrets.

G. B. Halt, Jr. et al., *Intellectual Property in Consumer Electronics,*
Software and Technology Startups, DOI: 10.1007/978-1-4614-7912-3_8,
© Springer Science+Business Media New York 2014

disclose to the public how to make and use the invention. Thus, even a competitor who independently, and without any knowledge, develops or reverse engineers an invention that is covered by a patent cannot practice the invention without infringing the patent. In contrast, with a trade secret, if a person that is privy to the trade secret unlawfully uses or discloses the trade secret, its owner can enforce the trade secret by filing a suit for misappropriation. On the other hand, if a person independently discovers or develops that particular trade secret, independently of the "first" trade secret owner, the "first" trade secret owner has no recourse. Once that trade secret is disclosed to the public, it is inevitably lost. Trade secrets are litigated less frequently than patents because the owner may not want to disclose the trade secret as part of the litigation discovery process or in court proceedings.

8.3 Right Creation and Term of Protection

A third consideration concerns the process for creating rights and the term of those rights. Unlike patents, there are no formal application or registration requirements for trade secrets. Any valuable and secret information used by a business is protected, as long as the business takes reasonable steps to keep it secret. This generally means the initial cost of trade secrets is lower. However, sometimes the long-term costs of enforcing and updating internal corporate procedures to keep information secret is more than the cost to secure a patent.

The following table summarizes these and other differences between patents and trade secrets.

Table 8.1 Patent and trade secret differences

	Patent	Trade Secret
What is the protected subject matter?	Inventions (e.g., processes, machines, manufactures, compositions of matter, improvements of the foregoing, etc.).	Business information which gives the owner a competitive advantage over competitors and is maintained in secret (e.g., formulas, patterns, compilations, mailing lists, programs, devices, methods, techniques, processes, etc.).
What is the term?	20 years from patent application filing date.	Indefinite, as long as information is kept secret and used in the business.
How is it acquired?	Filing a patent application with the USPTO.	Acquired upon creation. No formal application process.
What are the requirements?	The invention must be patentable subject matter, useful, novel, and non-obvious.	The information must give the owner a competitive advantage and must be maintained as a secret.

(continued)

Table 8.1 (continued)

	Patent	Trade Secret
Anticipated costs?	Patent application filing fee, patent issue fee, post-allowance maintenance fees, and attorney time required to prepare and prosecute a patent application.	No specific costs. However, costs are typically incurred in trying to maintain the subject matter as a secret (i.e., confidentiality and non-disclosure agreements, implementing internal policies for treatment of confidential information, etc.).
Can others use the invention/ information?	Others cannot practice the claimed invention without permission (i.e., license). However, others may design around the invention.	Trade secrets can lawfully be reverse engineered by others. In addition, others may use the information pursuant to a non-disclosure or confidential agreement.
Can the rights be lost?	Patent rights can be terminated if the validity of the patent is challenged. In addition, rights to a patent can expire if the invention was in public use or sold or offered for sale more than one year prior to filing a patent application (typically through the inventors own actions).	A trade secret is extinguished if it is disclosed to the public. For example, if an owner of a trade secret files a patent application for the invention, the trade secret is lost upon publication of the patent application or issuance of the patent.
How is it enforced?	Assert rights against others for patent infringement.	Assert rights against others for misappropriation of trade secret.

Patent protection is typically favored when:

- It is likely that a product can be reverse-engineered;
- The innovation might be discovered by others simultaneously;
- The technology is difficult or costly to keep secret;
- The technology must be disclosed to be of use;
- The subject matter is patentable; and
- The commercial value of the innovation exceeds the registration and maintenance costs.

Trade secret protection is typically favored when:

- The subject matter may not be patentable;
- The subject matter is part of a relatively "crowded" art;
- Keeping the innovation a secret is realistic and would not place an undue burden on the corporation;
- The potential market is likely to last longer than 20 years;
- The technology is developing rapidly and the innovation is likely to be obsolete in a few years.

8.4 Case Study: Trade Secrets and Cloud Computing

Cloud computing is a very popular tool used by many businesses to manage their information systems. Cloud computing is a strategic technology designed to deliver computing and storage capacity as a service that is provided by a third party vendor. A business and its employees are the end users of cloud computing services. All of the storage space and software used by end users is stored remotely on servers that are owned and maintained by the third party. The end users typically access the services and storage space of the cloud in exchange for a fee. Cloud computing is an attractive and cost-effective alternative to managing and maintaining information systems in-house. While moving business operations to the cloud may be more convenient and efficient than traditional means, there are some potential risks to be considered before making the business decision to move to the cloud.

Because trade secrets are only protected by the fact that they are secret, and protection is forever lost once disclosed. Efforts to protect trade secrets should be vigilant and cautious. It is never a good assumption that information uploaded to the cloud will be safe without first learning about the safety mechanisms put in place by the cloud vendor. Does the cloud vendor practice data encryption on its servers? What sort of anti-hack precautions are taken by the cloud vendor? Will the cloud provider submit to security audits? Additionally, the business should be aware that there is no guarantee that a hack or inadvertent leak of its confidential information will not occur. Prospective end user companies that are interested in moving their business to the cloud should first investigate the technical details of how the cloud provider protects client information.

Whether a company's trade secret information, which is stored in a cloud, is likely to lose its trade secret status as a result of being stored in the cloud (by a third party) is a legal question that is yet to be determined by the courts. Because the trade secret owner must take reasonable efforts to maintain the secrecy of the trade secret, the cloud vendor, acting as the agent for the company, must also take reasonable efforts to maintain its end user clients' trade secret information. In order to best protect the interests of the company, the company should investigate in detail the reasonable efforts claimed to be taken by the cloud vendor for protecting the information of its end user clients.

As cloud computing has grown, so have instances of cloud hacking and the theft of important business data. For example, LinkedIn, the business-oriented social networking site, experienced a breach of security in June of 2012 when its password database was hacked. It is estimated that 6.5 million LinkedIn passwords were stolen as a result of the breach. To make matters worse, a spam campaign spawned shortly after the breach that utilized service messages pretending to be from LinkedIn. LinkedIn members would receive the email, and concerned that their password had been exposed, unsuspecting email recipients would clink the link provided in the email and be bombarded with spam for a menagerie of pharmaceuticals.

Similarly, Sony's PlayStation network suffered a loss of customer data when its network and cloud music subscription service, Qriocity, were hacked in April of 2011. Personal data was stolen from over 70 million customers. The scope of the breach was so large that PlayStation had to shut down its network and go offline for more than a week to perform forensic analysis. Customers lost access to their accounts, could not play their video games or access music stored on the network and cloud.

Just like personal customer information can be stolen when the computing cloud is hacked, business information and trade secrets can also be stolen or misappropriated. For trade secrets that are business critical, or crown jewels of the business, it may be best to not even subject such valuable, confidential information to the risks associated with cloud usage. Once it is gone, it is gone for good.

Chapter 9
Intellectual Property Strategies for Software: Patent and Copyright Protection

9.1 Protecting Software Intellectual Property Rights

Patent protection and copyright protection for software are not mutually exclusive. Both can be obtained simultaneously because whomever creates the code is both the author (for copyright protection purposes) and the inventor (for patent protection purposes) of the software. As previously mentioned in earlier chapters, copyright protection only covers the expression of the code, meaning only the "word for word" copying of the code is protectable under copyright. Copyright does not protect the overarching idea behind the innovative code, which often times is where the true commercial value of the software lies. Thus, there is a need for patenting innovative software ideas. Patents protect the particular idea, or invention, of the software.

The following table summarizes these and other differences between patents and copyrights.

Table 9.1 The differences between patent and copyright protection

	Patent	Copyright
What is the protected subject matter?	Inventions (e.g., processes, machines, manufactures, compositions of matter, improvements of the foregoing, etc.)	Literary works; musical works, including any accompanying words; dramatic works, including any accompanying music; pantomimes and choreographic works; pictorial, graphic, and sculptural works; motion pictures and other audiovisual works; sound recordings; and architectural works
What is the term?	20 years from patent application filing date	Life of the author, plus 70 years
How is it acquired?	Filing a patent application with the USPTO	Registration with the copyright office

(continued)

G. B. Halt, Jr. et al., *Intellectual Property in Consumer Electronics, Software and Technology Startups*, DOI: 10.1007/978-1-4614-7912-3_9, © Springer Science+Business Media New York 2014

Table 9.1 (continued)

	Patent	Copyright
What are the requirements?	The invention must be patentable subject matter, useful, novel, and non-obvious	Original works of authorship fixed in any tangible medium of expression
Anticipated costs?	Patent application filing fee, patent issue fee, post-allowance maintenance fees, and attorney time required to prepare and prosecute a patent application	Copyright protection is free. Registration of the copyright has an associated fee. There can also be other special services fees
Can others use the invention/ copyrighted material?	Others cannot practice the claimed invention without permission (i.e., license). However, others may design around the invention	Others cannot use the copyrighted material without permission (i.e. license). However, others can find another way to achieve the same result that falls outside of the scope of the copyright
Can the rights be lost?	Patent rights can be terminated if the validity of the patent is challenged. In addition, rights to a patent can expire if the invention was in public use or sold or offered for sale more than one year prior to filing a patent application (typically through the inventors own actions)	Copyright is created in a work once it is fixed into a tangible medium of expression. Registration is not required to obtain copyright. Registering does provide extra protections for the copyright holder. Rights expire with termination of the copyright
How is it enforced?	Assert rights against others for patent infringement	Assert rights against others for copyright infringement

9.2 Patent Protection for Software

Patent protection is available for software and is becoming a more popular means of protecting the intellectual property behind software, particularly for small startup software companies. It is understandable that a startup with limited capital may view the patenting of its software to be a waste of precious funds. In an industry like software programming, where code mutates and develops at incredible speed, it may seem like the pace at which a patent makes its way through the United States Patent and Trademark Office is that of a snail. This perception often leads startup software companies to the presumption that it is more affordable and a better business strategy for the startup to utilize first-mover advantage to monetize newly developed software, rather than bother with obtaining a patent on the software.

While first-mover advantage has the benefit of turning the company's newest software development into cash quickly, the risk associated with utilizing first-mover advantage as a business strategy is high and can doom an up-and-coming company once its competitors get a hold of its unprotected software. If the startup has developed software that spurs the development of a new technology, or serves as a base platform for the development of new technology, the implications of being the owner of that software could be immensely profitable if the IP rights are properly protected.

As the use of software has grown, for patentability purposes, software-based inventions have been characterized as a collection of processes, or when in conjunction with a computer, it has been considered a unique machine. There are instances where software has been considered to fall into both categories as well. To be patentable subject matter, it is required that the involvement of a computer or the Internet be integral in the software method's successful operation.[1] The software must be non-obvious and useful.

Sometimes software consists of additional elements which may be in need of patent protection, such as the Graphical User Interface (GUI) associated with a program. These nonfunctional and ornamental aspects of the program may be protected with a design patent. "Whoever invents any new, original, and ornamental design for an article of manufacture may obtain a patent therefor."[2] An object with a design that is substantially similar to the design claimed in a design patent cannot be made, used, copied or imported into the United States. Design patents are limited in their scope, protecting only ornamental, non-functional designs rather than functional elements. Protection under a design patent is shorter than that for a utility patent, and lasts for only 14 years from the date of issuance, and the coverage of the design patent is defined by a single claim directed to the images included in the design patent. Any elements of the design patent drawings which are drawn in dashed lines represent an unclaimed feature of the design, meaning it is not part of the patented design.

Design patents can be sought for many different ornamental designs. Some examples include the design of a product, product packaging, font, icons and patterns. Design patents provide brand protection because they protect the appearance of a product or its packaging, preventing another from creating a copy or look-a-like product. Due to the narrow scope of design patents, it is a common intellectual portfolio management strategy to procure multiple design patents for a single product. Each design patent embodies a slight variation or modification to the original design. Possessing a collection of design patents on a single product broadens the scope of protection afforded to the design of the product.

In recent years, GUIs have been the subject of design patents in the United States. GUIs are eligible for design patent protection if the user interface is embodied on a display screen; that is, the drawings in the design patent depict an icon or interface embodied in an article of manufacture, such as a computer screen, smartphone display panel, or monitor.

> *Example*: U.S. Design Patent D608,366 claims "The ornamental design for a graphical user interface for a display screen or portion thereof, as shown and described." The patent also includes a Figure depicting the claimed GUI.

---

[1] Ultramercial, LLC v. Hulu, LLC, 657 F.3d 1323 (Fed. Cir. 2011), *reh'g and reh'g en banc denied*, No. 2010-1544, 2011 U.S. App. LEXIS 25055 (Fed. Cir. 2011).

[2] 35 U.S.C § 171.

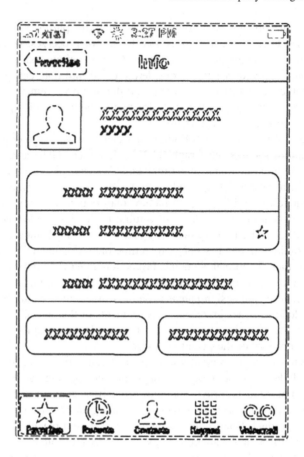

GUI embodiments drawn in an article of manufacture satisfy the "article of manufacture" requirement of 35 U.S.C § 171. Furthermore, non-static computer generated icons, or icons that change in appearance during viewing, may also be protected by a design patent. The change must be noted in the claim language and shown in two or more views in the design patent drawings. The drawings should depict the change in the icon over time.

Example: U.S. Design Patent D613,300 claims "The ornamental design for an animated graphical user interface for a display screen or portion thereof, as shown and described." The patent also includes 18 Figures (10 of which are shown) depicting how the claimed computer generated icon changes over time.

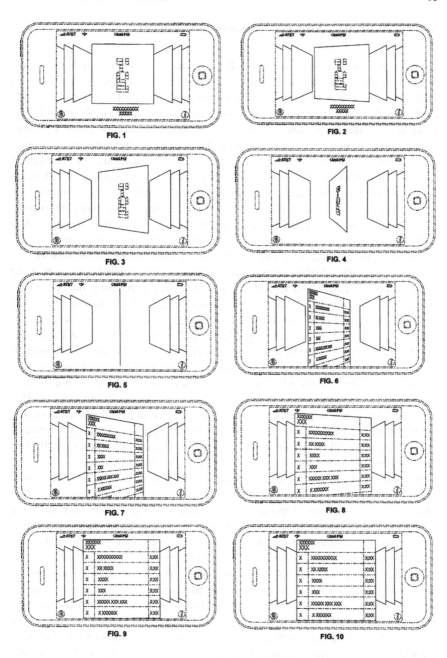

FIG. 1   FIG. 2

FIG. 3   FIG. 4

FIG. 5   FIG. 6

FIG. 7   FIG. 8

FIG. 9   FIG. 10

9.3 Case Study: Amazon's One-Click Patent

Everyone is familiar with making online purchases: selecting an item to buy, placing it in his or her digital shopping cart, and then proceeding to check out. At check out, an individual enters their personal purchasing information, i.e., credit card and billing information. Anyone who has engaged in online shopping is also familiar with the frustrations and occasional inconvenience of using an online shopping cart to make an online purchase. For example, after entering all of your personal information, you accidentally strike a key by mistake and it all disappears forcing you to reenter all of your information. Sometimes shoppers will abandon their cart, because they are confused as to how to complete the purchase, or because they are frustrated with the online buying system. They may plan to come back to their cart at a later time to complete the transaction but ultimately they forget about it and never complete their online purchase. Abandoned digital shopping carts are effectively lost Internet-based sales for the online retailer.

To help capture these lost sales, Amazon.com sought a patent on a way to make online shopping easier. Amazon developed an online purchasing method where shoppers could complete a buying transaction with a single mouse click. The United States Patent Office issued U.S. Patent Number 5,960,411 (the '411 patent), entitled "A Method and System for Placing a Purchase Order via a Communications Network," to Amazon.com on September 28, 1999; the '411 patent is commonly referred to as the Amazon "1-Click" patent. The '411 patent describes online purchasing system where customers enter their credit card number and billing address information on their initial visit to the Amazon.com website. This personal information is stored by Amazon's database system and when a customer returns to the website at a later point in time to make purchases, the customer will not have to reenter their personal information to make an online purchase. Rather, the customer can make their purchase with a single click of the mouse.

Around the same time that the Amazon "1-Click" patent issued, Barnes & Noble's online shopping website was using an online purchasing system called "Express Lane." Express Lane allowed customers to pre-register their personal information and make purchases directly with a single click of the mouse, rather than requiring customers to go through the traditional online purchasing process where a customer places an item in a digital shopping cart and then "checks-out" the contents of their cart. Express Lane involved the use of a cookie on the customer's computer which would signal to the Barnes & Noble website that the buyer was a pre-registered customer and that their personal information has been stored on Barnes & Nobel's servers. Unfortunately for Barnes & Nobel, the Amazon '411 patent also utilized a cookie in order to complete its 1-click purchasing experience. Amazon had the patent protection and Barnes & Nobel did not.

Amazon won injunctive relief from a federal district court in Washington, a judgment from which Barnes & Nobel appealed. The Court of Appeals affirmed the injunction and Barnes & Noble was forced to shut down their Express Lane one click shopping check out system. Barnes & Nobel was required to resort to the traditional shopping cart method of online transactions. After much back and forth between the two online retailers, a settlement was negotiated.

The Amazon 1-click patent has generated a substantial amount of debate over whether software applications should be patentable subject matter. The arguments for software patentability are readily apparent in the Amazon example. By being the only entity holding a patent for completing online transactions in a single click of the mouse, Amazon economically benefits by excluding others from using its patented method. One of the consequences of allowing software to be patentable subject matter is that the Amazon 1-click patent reduces efficiency in online purchasing. Other online retailers, both those that directly compete with Amazon, and those that do not, are unable to use a one-click purchasing model, giving Amazon a monopoly on the technology. It has also been contended that the technology is too obvious to warrant patent protection.

9.4 Copyright Protection for Software

A copyright is created in a work once it is fixed into a tangible medium of expression. For example, a programmer creates copyright protection in the code he or she is developing the moment he or she hits the "save" button. Obtaining a copyright is essentially "free" since registration of the copyright is not required, but is highly recommended.

As part of the registration process of the code, the author must provide the Copyright Office with a copy of the identifiable portions of the source code, that is, the first 25 pages of code and the last 25 pages of code. If the code is less than 50 pages, then the author must submit the entire code. If portions of the code are to be protected as a trade secret, the code may be submitted with a letter stating that it is a trade secret. In which case, the author must submit the first 25 pages of code and the last 25 pages of code, but the trade secret portions may be blocked out. Both the source code and the object code are copyrightable.

9.5 Case Study: The Litigation Between Oracle and Google

In 2010, Oracle initiated a lawsuit in U.S. District Court for the Northern District of California against Google regarding Google's use of the Java programming language in its Android mobile operating system. Java was initially created by Sun Microsystems, which was purchased by Oracle in 2010. Based on Sun's Java language, Oracle asserted two main complaints: (1) Google's use of Java in Android is a violation of Oracle's software patents, and (2) Google's use of Java language and use of 37 application programming interfaces (APIs) was a violation of Oracle's copyrights. What is interesting about this litigation is that the outcome

Example of trade secreted programming code blocked out.

```
\\use 50% one-shot finish, updated to 1.5 cool down after
patch on 10/28 DC
    do {
        int            actionA, actionB;
        if (                                              )
            {
                    actionA=1;
            }
    else
        {
        if (cd >1.5)
                {
                at=1;
                cd=T-1.5;
                }
        else
                {
                at=2;
                cd=T=0.5;
                }
        }
    else
    { ...
```

could impact whether APIs are subject to copyright protection. Typically APIs are denied copyright protection under US copyright law as they are functional in nature; they are used to promote interoperability since the APIs are designed to enable a vendor's software to interact with the programming language; in this case Java. The case was separated into three sections; the patent portion of the suit, the copyright portion of the suit, and damages.

9.5.1 Oracle's Patent Suit

When Oracle first came to court with its complaint, it asserted seven software patents related to Java. However, prior to trial, Google sought reexamination from the USPTO of all the patents asserted by Oracle. The USPTO invalidated five of the patents, and many of the claims of the remaining two. What ultimately ended up before the jury was evidence of infringement on eight claims from the two remaining patents[3] pertaining to virtual machine software memory management and performance. The jury was unconvinced by the evidence and found that Google did not infringe the two Oracle patents.

---

[3] US Patent No. 6,061,520 and US Reissued Patent No 38,104.

9.5.2 Oracle's Copyright Suit

Oracle also asserted that Google violated Oracle's copyright in Java and 37 APIs which it acquired the IP rights to from Sun Microsystems in 2010. Google made the argument that Java was open-sourced by Sun Microsystems in November of 2006 and therefore use of the language was not copyright infringement. By open-sourcing Java under the General Public License version 2 (GPLv2), a free software license, Sun had made its Java code available to anyone who sought to use it. In terms of not copying Oracle's APIs, Google used a substitute, independently developed, open-source programming and development tools. Furthermore, Google contended that APIs are not subject to copyright protection.

At the time the jury gave its verdict on the patent infringement issue, it also gave an incomplete verdict on the copyright issue, finding that Google did copy nine lines of code from Java (out of millions), but Google removed the code from its Android operating system, thereby limiting Oracle's recovery to only statutory damages; a maximum of $150,000. The jury's verdict was incomplete because the judge had yet to make a determination if APIs are copyrightable, as a matter of law. Upon review of the matter, the judge concluded that APIs are not copyrightable under the Copyright Act. Since APIs are small sections of code that function as "how to" instructions for the rest of the program, they are functional as methods of operation.

9.6 Business Method Patents that Protect Software

Business method patents are a group of patents that claim and disclose novel methods for doing business. Business method patents are often considered eligible for patentability because they are categorized as "processes." Typically, a business method patent incorporates some sort of electronic commerce (e-commerce) or digital means of doing business, and often incorporates the use of a software program.

In 2010, the United States Supreme Court, in *Bilski v. Kappos*,[4] affirmed that business methods are patentable so long as the claims fall within "process" subject matter of the Patent Act. The holding of *Bilski* can be extended to software patent eligibility since software is often categorized as a protectable process. To evaluate whether a process is patentable, it must be analyzed under the machine or transformation test. The Court indicated that the "machine and transformation test"—a test of patent eligibility for process claims, where the process claim (1) is implemented with a particular machine ...one specifically devised and adapted to carry out the process in a way that is not concededly conventional and is not trivial; or else (2) transforms an article from one thing or state to another—was not the exclusive

---

[4]   130 S. Ct. 3218 (2010).

test for assessing the patent-eligibility of process claims. The Court stated that it was "unaware of any ordinary, contemporary, common meaning of 'process' that would require it to be tied to a machine or the transformation of an article,"[5] supporting the premise that processes, for the purposes of patentability, can consist of intangible elements and still be patentable. Software, both in general and as part of a business method, often describes a process that does not consist of tangible elements.

To satisfy the machine prong of the test, there must be the use of a specific machine that imposes meaningful limitations on the scope of the patent claims making them patent eligible. On the other hand if the process is relying on the transformation prong of the test to determine patent eligibility, the process must transform a particular article to a different state or thing. An article includes a physical object. Transformation of the article must change the article to a different state; i.e., the transformation can be particularly identified. A new or different function or use can be evidence that an article has been transformed. The transformation of the article must be particular, i.e., the method must involve a specific article such that the transformation imposes real world limits on the claimed method by limiting the claim scope to a particular practical application.

The USPTO issued a quick reference sheet listing factors to be considered by patent examiners when making a patent eligibility determination post-*Bilski*.[6] Affirmative answers to the following questions tend to lend weight to the categorization of the claimed matter, either patent eligible or not. Some of the considerations that weigh in favor of being categorized as patent eligible subject matter include:

- If there is a recitation of a machine or transformation (either express or inherent) in the claim:

  - Is the machine or transformation particular?
  - Does the machine or transformation place meaningfully limitations on the execution of the steps recited in the claim?
  - Does a machine implement the claimed steps?
  - Is the article being transformed particular in some way?
  - Does the article undergo a change in state?

Is there an objectively different function or use for the article after the transformation?

  - Is the article being transformed an object or substance?

- If the claim is directed toward applying a law of nature:

  - Is a law of nature practically applied?
  - Does the application of the law of nature meaningfully place limitations on the execution of the steps recited in the claim?

---

[5] Id. at 3221.

[6] See generally http://www.uspto.gov/patents/announce/bilski_qrs.pdf.

- If the claim is a statement of a concept:

  - Is the claim more than a mere statement of a concept?
  - Is the claim directed at a particular solution to a problem that needs solving?
  - Does the claim implement a concept in a tangible way?
  - Can the steps of the performance be observed and verified?

  Some of the considerations that weigh against being categorized as patent eligible subject matter include:

- If the recitation of a machine or transformation (either express or inherent) is lacking from the claim.
- If there is insufficient recitation of a machine or transformation:

  - Is the involvement of a machine or transformation merely nominal, insignificant, or tangential related to the performance of the claim steps?

    Is the machine merely performing data gathering operations?
    Is the claim merely indicating that use of the method is limited to a particular field?

  - Is the recitation of the claimed machine so generic that it could cover any machine capable of performing the claimed step or steps?
  - Is the machine simply an object on which the claimed method operates?
  - Does the transformation only involve a change in position or physical location of article?
  - Is the "article" merely a general concept? General concepts being things like

    Aspects of human behavior
    Fundamental economic principles or practices
    Fundamental legal theories
    Mathematical concepts
    Mental exercises
    Teaching models

- If the claim is a mere statement of a general concept:

  - Would a grant of the claim give the patentee a monopoly on the use of that concept as described in the claimed method?
  - Are both known and unknown uses of the concept covered by the claim, and could it be performed through any existing or future-devised means?
  - Does the claim lack a solution; only expressing a problem to be solved?
  - Are the means by which the claimed steps are applied subjective or imperceptible?

- If the claim is not directed at the application of a law of nature.

  - Would a grant of the claim give the patentee a monopoly on a natural force or scientific fact?

- Is the application of the law of nature merely a subjective determination?
- Is the involvement of the law of nature merely nominal, insignificant, or tangential as it relates to the performance of the claimed step or steps.

Chapter 10
Developing and Managing an Intellectual Property Portfolio

It is important to have a well organized and focused intellectual property management program in place in order to properly develop and enforce a company's intellectual property rights. Factors that should be considered when developing an intellectual property portfolio management program include: (1) strategic considerations in developing an IP portfolio; (2) administrative issues associated with managing the IP portfolio; and (3) ongoing IP diligence protecting rights and pursuing others.

This chapter will describe a systematic approach how to develop and actively manage an IP portfolio. This chapter will also explore as a case study the IP strategy employed by Fujitsu LTD, a Japanese information technology equipment and services provider. Fujitsu has a demonstrated commitment to developing and managing its IP portfolio. The company's IP strategy is very well organized and deeply integrated into the company's overall business strategy. The company emphasizes this importance by explicitly referencing its commitment to IP strategy in the company's code of conduct and business philosophy. Employees are educated about company IP policies and are incentivized by an inventor compensation system to invent new products. The company also has a regular practice of creating internal start-up companies and spin-offs based upon employee inventions. Fujitsu implements a global strategy, filing domestically and abroad (protection is primarily sought in Europe, other parts of Asia, and North America).

10.1 Developing an Intellectual Property Portfolio Strategy that Fits Corporate Strategy

Developing a company's IP portfolio strategy requires 5 steps: (1) identifying the corporate strategy and determining how best to align the corporate strategy with the IP strategy, (2) identification of existing IP assets (an IP audit); (3) determining which of those assets are "core" assets (i.e., those assets that have a strategic importance to the company); (4) allocating resources to core and non-core assets as

G. B. Halt, Jr. et al., *Intellectual Property in Consumer Electronics,*
Software and Technology Startups, DOI: 10.1007/978-1-4614-7912-3_10,
© Springer Science+Business Media New York 2014

appropriate; and (5) setting up a program to ensure that those assets are periodically reviewed and maintained to ensure that the IP portfolio is developed consistently with the corporate business plan.

10.1.1 Identifying the Corporate Strategy and Determining the Best Means for Aligning the Corporate Strategy with the IP Strategy

The first step to developing a company's IP strategy is to start with identifying the
corporate business strategy. Once there is an in depth understanding of what drives
the company, it is easier to proceed through the next four steps for developing a
company's IP portfolio strategy in a way that is highly coupled to the overall
business strategy and objectives of the company.

10.1.2 Identification of IP Assets

In order to properly identify IP assets, it is critical that a thorough IP audit be
performed by personnel within the corporation who are knowledgeable or familiar
with the company's IP or the company's technology plan. This should also be
performed in coordination with an experienced IP attorney. The IP audit should
identify copyrights, trademarks, trade dress, trade secrets, confidential information,
copyrights, mask works, domain names, industrial designs, patents, patentable
inventions, and other IP assets owned by the company. Some registered assets (for
example, patents, trademarks and copyrights), require periodic maintenance in the
form of payment of official fees and filing of official documents. If these fees are
not paid or the official documents are not timely filed, these rights can expire.
Consolidating all of this information in a single location can help ensure that all of
a company's intellectual property can be easily identified, tracked, and maintained.

Along with intangible assets, an intellectual property audit should identify any
encumbrances on the company's IP assets. These encumbrances include agreements
(such as distribution agreements, licenses, software licenses, franchise agreements,
assignments, covenants not to compete, employment agreements, third party
development agreements, and liens against the intangible assets) and any other
liabilities that a company may have with third parties. The audit should identify
whether or not these agreements affect the company's IP or other intangible assets.

Example: Fujitsu's patent portfolio contains over 100,000 issued patents and
patent applications from around the world. The patent portfolio is broken
down into four board technology segments, which represent the specific core

areas of industry in which Fujitsu competes. The breakdown is as follows: common technologies and future business related IP—this segment includes technology related to laboratory equipment and corporate center technology; electronic devices, including electronic circuit technology; ubiquitous product solutions; and technology solutions, including communications related technology and information processing technology.

10.1.3 Determining Whether the Identified IP Assets are Core Assets

Following IP asset identification, the assets must be evaluated against the company's business plan and current financial condition to determine whether the assets are core assets. A core asset is one that is critical to the success of the company. This is important because IP assets can be very expensive to secure and maintain. The money spent on IP may have been diverted from other programs that are crutial to the company. Thus, IP strategy is often a balance between short- and long-term strategies and a company's economic resources.

The following factors help to determine whether IP assets are core assets and important to the success of a company:

- Are the IP assets (or the underlying technology or programs) currently used and will they be used in the future, or are they related to a product or a part of the business that is obsolete?
- Do the IP assets support a program (either marketing, legal or research and development) that provides the corporation with a competitive advantage?
- If the IP assets were lost, would the company be negatively impacted?
- Do competitors have IP assets that will negatively impact the company in the marketplace?
- Does the company need additional IP assets in order to cover the important aspects of the company's business plan?

Example: As with most technology-related businesses, Fujitsu's business is always changing. In order to be responsive to the market, Fujitsu periodically assesses whether its IP assets are still in alignment with the business goals of the company. It is important to Fujitsu that the company's IP generate income, so the company pursues means for capitalizing on those assets that are no longer related to the core operation of the company. The technology therefore is licensed to another entity, or the non-core asset is divested through a technology sale.

10.1.4 Properly Allocating Corporate Resources to Core and Non-Core IP Assets

Because there are many different types of IP assets, it is difficult to generalize the scope of protection for each type of IP. However, each type of IP typically has multiple levels of protection and these multiple levels of protection, in turn, have varying costs. Core assets should be given the broadest scope of protection since they are critically important to the corporation. Non-core assets may be given minimal protection; or even none at all. It is important to note that corporate resources include not only financial resources, but also human resources and executive attention.

For example, it could cost ten thousand dollars or more to secure a utility patent in the U.S. In addition, if there is a potential to use or sell the invention on an international level, patent protection may be required in individual foreign countries. Patent protection in foreign jurisdictions can be more expensive than the U.S. due to higher official fees and expensive translation costs. For example, filing a European or Japanese patent application can be nearly twice as expensive as a corresponding U.S. filing. One way to mitigate costs may be to file regular (non-provisional) utility patent applications for core IP assets and provisional patent applications for non-core IP assets. As described in Chap. 1 regarding patents, provisional patent applications provide a vehicle for a company to keep its options open for a 1 year period to decide whether to pursue a regular patent application. At the cost of approximately $1000–$3000 (depending on the complexity), this can be a minimal financial commitment for a company, although more costs will be incurred if the company decides to file a regular (non-provisional) patent application.

The degree to which corporate resources are devoted to certain IP assets depends largely upon the value of the IP to the company. If those assets provide minimal value, then a company may choose not to put a large amount of resources into defending the IP. An IP attorney can help correlate core IP assets with a company's goals in order to properly allocate the appropriate resources.

10.1.5 Setting up a Program for Periodic Review of IP Assets

Following the initial IP asset review, it is important to ensure that those assets are periodically reviewed. These successive reviews ensure that existing IP assets are properly maintained, and an informed decision is made to retain, sell, or dispose of non-used assets.

The frequency at which such reviews are conducted will vary greatly, depending on whether the company is a startup, an emerging entity or a long established corporation. At the very least, a yearly audit should be conducted. Preferably, quarterly reviews of a current portfolio should be undertaken. The amount of resources to devote to the review will also depend upon the importance of the IP assets to the company. If the only core asset that has been identified is the corporate name and that has been protected with a trademark registration, in-depth IP audits on a yearly basis may not be necessary.

Example: Fujitsu routinely reviews its IP for opportunities to enhance its portfolio. For instance, pending patent applications are periodically reviewed to determine if there is any content which can be protected via supplemental or divisional patent application procedures, or whether protection should be sought in additional countries. Non-core assets may also be dropped.

10.2 Administrative Issues for Long-Term IP Portfolio Management

After building an IP portfolio, administering the IP portfolio is a responsibility that requires integration of knowledge from many different parts of an organization. A well-executed IP portfolio management program must be closely tied to a company's strategic plan, marketing initiative, legal department, and corporate research and development initiatives. Although tying all these aspects together in a comprehensive IP portfolio management program may appear to be daunting, there are a number of steps that make the management of an IP portfolio a much easier endeavor, regardless of the size of the IP portfolio.

10.2.1 Docket Management

Obtaining and maintaining intellectual property rights is a deadline-driven process. If a company does not meet its IP-related deadlines, the consequences can range between payment of late fees or penalties to a complete loss of the IP rights. Accurately tracking U.S. and international deadlines ensures that formal requirements are fulfilled. Such tracking is the primary responsibility of an IP portfolio docketing system maintained by the company or its IP counsel.

Docketing software is specifically tailored for IP portfolio management to intake relevant information required to properly track IP deadlines. The software elicits required information and calculates the relevant deadlines for both U.S.

cases and all foreign counterparts. The software also provides reminders at certain intervals as the deadlines approach, and these reminders can be automatically e-mailed to the appropriate individuals. Such reminders can be tailored to provide not only information regarding the particular IP asset, but also information related to an organization's procedure.

For example, rather than taking the information generated by the docketing software, inserting this information into a memo and sending it to the person within the organization who will be responsible for making the decision regarding whether or not to proceed with complying with a deadline (which may include the payment of significant fees), automatic notices, which include the IP asset at issue as well as the specific steps that are required to be taken within the organization, can be automatically provided by IP docketing software. Thus, the software can be specifically tailored to the procedures of an organization in order to make management of the IP portfolio more efficient. If an organization or IP portfolio is not large enough to justify the purchase of such specialized software, this service may be contracted to third parties.

10.3 Ongoing IP Diligence: Protecting Rights and Pursuing Others

10.3.1 Defending Your IP

Aggressively pursuing competitors with litigation can be reckless if the company is vulnerable to an easy counterattack. Pursuing a solid defensive strategy means putting into place the minimum measures in order to protect core IP assets.

IP protection is, by its very nature, defensive. Patents prohibit others from using the products or processes generated from research and development programs. For a technology-centric company, this can represent the core of the business. Trademarks and service marks protect one of the most important aspects of a corporation's image; its name, logos, etc. Significant amounts of marketing and advertising budgets are typically allocated to using trademarks and service marks and, therefore, it is extremely important to protect these valuable assets. Copyrights protect a corporation's expression of ideas such as drawings, websites, marketing materials, documents, website designs and software. Trade secrets, if handled properly, help protect customer lists, business plans and strategic plans. Since large amounts of a company's resources are typically allocated to all of these endeavors, it is important to consider protection for all aspects of a company's IP portfolio.

Pursuing a vigorous defensive strategy has additional benefits. Such a strategy not only insures that the IP rights are protected, but can also strengthen those IP rights. For example, with effective trademark protection, if a trademark holder does not provide notice to competitors or the public at large from using a trademark to refer to a particular product (for example, using the term "Kleenex" as opposed to "facial tissue"), the trademark may ultimately become "genericized," whereby anyone can use the mark. Thus, rights to the mark will be lost and it will cease to be an asset for the company. This happened with respect to the term "aspirin" for the Bayer Corporation and the term "escalator" for the Otis Elevator Company.

A similar defensive strategy may be pursued using patent protection. If a corporation has a product that is protected by a patent, it can defend against another company trying to copy that product by asserting the patent against the infringer. If successfully implemented, such a strategy will protect the product and, as a result, the company's market share.

There are many different reasons why companies initiate intellectual property litigation. One of the primary defensive weapons that a company may have is to counter a claim of infringement with a counter attack using its own IP portfolio. When a company that is aggressively pursuing patent infringement litigation realizes that the alleged infringer is not going to "cave in," the company may become more reasonable regarding its demands. Without a defensive patent portfolio as a counter-threat, a company can become vulnerable to repeated attacks from different competitors. Often, entire industries will be cross-licensed in such a fashion. However, such cross-licenses are only granted to those companies that have an IP portfolio to bring to the table.

Several measures that should be considered when putting together a strong defense include:

Are all core assets protected?

It is critically important to ensure that all IP assets that have been identified as core assets are properly protected. If they are not protected, a company may be vulnerable to a competitor entering the market with a similar competing product, thereby quickly eroding the company's market share or operating margins.

Having IP assets on fundamental technology will help insulate a company from attacks by competitors having IP assets on related technology. Although such claims may be frivolous, defending against these attacks can drain smaller companies of desperately needed capital and distract management from its primary focus.

Have clearance opinions been sought on all products to ensure that a company is not charged with infringement of competitors' IP portfolios; particularly patents or trademarks?

Just as critical as protecting a company's core assets, is ensuring that its products do not infringe competitor's IP rights. One way to determine this is to periodically review the industry for competitors' IP assets. If relevant IP assets are identified, it is recommended that a company seek a clearance opinion or a freedom to operate opinion from an experienced IP attorney. This can help ensure that the company's invention is free to be used or sold without charges of infringement by competitors.

Have vulnerabilities or gaps in competitor's product lines or IP assets been identified?

Once a company assesses its own vulnerabilities, it is critical to identify those of its competitors. Although this does not require immediate or offensive action, such information will become invaluable should a competitor begin to make demands of the company.

If you have determined not to pursue patent protection for an invention, have you considered a defensive publication in order to keep a competitor from gaining right to the idea?

As discussed above, a company will need to make strategic decisions about which IP rights to expend resources on for protection. If a decision is made to forego protection for a particular asset, a company should consider whether the asset could provide competitors with an advantage if independently developed. In order to keep competitors from gaining such a strategic advantage, publication of an article can dedicate information about the invention to the public in general and mitigate the harm resulting from a competitor's use of the invention.

10.3.2 Leveraging Your IP Rights

An IP portfolio is a significant strategic asset and can be used to leverage favorable outcomes against competitors. A primary offensive weapon is the threat of an infringement suit. A company can also use licensing and the threat of (or actual) litigation in order to generate revenue from its IP assets.

Example: Fujitsu understands the value of leveraging IP assets. In the early 2000's Fujitsu became involved in an infringement suit with Samsung over Fujitsu patents pertaining to basic technology for plasma display panels. The litigation came after Fujitsu had tried for years to negotiate compensation from Samsung. When negotiations did not produce the desired outcome, Fujitsu pursued remedies through the court system on two fronts. First, Fujitsu filed suit in Tokyo, Japan, asserting Japanese Patent No. 2,845,183 and seeking injunctive relief. Next, Fujitsu filed suit in U.S. Federal District Court in Los Angeles, California asserting 10 U.S. Patents. Samsung countersued, challenging the validity of Fujitsu's U.S. patents. The Japanese court granted Fujitsu's injunction, and within two months, the parties reached a settlement agreement. The resolution involved the execution of a cross-licensing agreement between the parties.

Licensing and cross-licensing are other techniques used by Fujitsu to leverage its IP. Fujitsu considers cross-licensing as a way to ensure business flexibility. Several large corporations have entered into cross-licensing agreements with Fujitsu, including Samsung Electronics, International Business Machines, Motorola, Texas Instruments, Hynix, ARM and Intel Corporation.

When deciding to aggressively pursue an offensive IP strategy, corporate management must be willing to back up any infringement allegations. There are several measures that should be considered when putting together a strong offense:

(1) Are competitors litigation averse (i.e., do they have a history of settling litigations quickly)?
(2) Is your corporate management and board of directors willing to use the threat of litigation (or actual litigation) in order to enforce the IP rights or gain a strategic advantage over your competitors?
(3) Is the company willing to monetize its unused IP assets through franchising, licensing, or sale?
(4) Is the company willing to develop IP assets solely for the purpose of monetizing them?
(5) Is the company willing to search for and purchase IP assets and other corporate assets from those which are available in the marketplace?

All of these measures should be considered carefully when crafting a winning offensive strategy. A company should carefully weigh the risks and benefits prior to embarking upon an IP strategy that incorporates an aggressive offense as a central part of its strategy. Such a strategy will only be successful if it has the full support of management and directors, as it will take up large portions of corporate resources.

Part III
Monetization of Intellectual Property Portfolios

Chapter 11
Intellectual Property Portfolio Acquisition

In addition to the traditional intangible corporate assets such as brand value, corporate reputation, franchises and human capital, IP assets (patents, trademarks, copyrights, and trade secrets) have emerged as important and valuable corporate assets. IP rights can be bought and sold like any other business commodity.

11.1 The Business of IP Portfolio Acquisition: General Valuation Guidelines and Due Diligence Concerns

One of the main challenges associated with IP portfolio acquisitions is that the value of the intangible assets contained within the portfolio is not fixed. Since the underlying assets are constantly changing, (i.e., aging, expiring, and new applications are being filed and examined) the value of a portfolio is constantly changing. There are a variety of factors which contribute to the value of an IP portfolio, such as market conditions and the business strategies of both the buyer and seller. Due diligence will be discussed in this chapter, as well as factors which contribute to the value of IP assets. For a more in depth discussion on IP valuation and valuation models, please refer to Chap. 12: IP Valuation.

Whether on the buying side of the transaction or the selling side, thoughtful portfolio management and portfolio asset optimization are important to realizing the value of a portfolio. Due diligence is intended to examine and analyze each asset contained within the portfolio, and to reveal the value of the portfolio assets by examining the strength, scope, enforceability, validity, ownership rights and future potential that may be derived from the IP assets. It generates a better understanding of the assets involved in the transaction and identifies any possible liabilities or encumbrances inherent in the rights.

A good rule of thumb when performing due diligence is to independently verify everything: ownership, interests, scope of rights, assignments, etc. The results of due diligence are often only as good as the effort, time and money that goes into it. Early IP due diligence can help the parties avoid costly mistakes and minimize the

G. B. Halt, Jr. et al., *Intellectual Property in Consumer Electronics, Software and Technology Startups*, DOI: 10.1007/978-1-4614-7912-3_11, © Springer Science+Business Media New York 2014

risk of potentially disappointing outcomes by enabling prospective purchasers to make an informed decision when making an offer. Perhaps due diligence indicated that a particular piece of IP is not worth as much as originally assumed. It would be appropriate in that case to negotiate a lower price, or possibly even back out of the acquisition altogether if necessary.

11.1.1 Seek Legal Advice

Seeking the advice of legal counsel will go a long way toward demystifying the IP due diligence process. Regardless of whether you are the prospective buyer of the portfolio, or the seller, it is important to procure legal assistance with the due diligence process. Legal counsel will assist in this process by providing guidance, documenting any appraisals or transactions and performing risk assessments and devising strategic solutions to any discovered or potential problems. It is important to assess whether the patents for sale have been involved in litigation and/or licensing agreements, and if so, what the outcome was. Legal counsel will also determine whether any foreign patent protection exists and how likely it is that the patents might be declared invalid or infringed and can also provide an assessment as to whether the patents are infringing other existing patents. Legal counsel will assess the IP portfolio in an unbiased fashion.

In order to better facilitate due diligence efficiently and thoroughly, legal counsel should be provided with a complete list of the intangible assets that are the subject of the acquisition. Counsel will use this list as a starting point for researching the IP rights to make a determination as to whether there are any other related assets that are missing from the list, and/or any issues that cannot be resolved by corrective action that would significantly reduce the value of the IP. Legal counsel will also need to be aware of any constraints that the buyer or seller may have, as well as any price expectations. If there are any time frame considerations, legal counsel should be informed of these as well.

11.1.2 Determine the Scope of the IP

It is important to define precisely what is to be bought or sold within the patent portfolio. The first step is to perform a thorough review of the IP if available in the portfolio. The portfolio should be organized in a way that makes logical sense. For example, technologically related IP should be grouped together.

The number of assets and the scope of protection those assets have is often the most important basis of value. To think of it another way, the identified goods and services of a registered trademark or the claims of a patent are analogous to the property lines of real estate. They define the boundaries and scope of the intellectual property asset. A prospective buyer will certainly want to know the specifics of the boundaries and scope of the intangible assets of an IP portfolio, particularly freedom-to-operate considerations, the validity of the IP, whether the IP is in force, proper ownership of the IP and the scope of protection afforded to the assets.

11.1.2.1 Freedom-to-operate

Freedom-to-operate refers to whether the buyer will be able to practice the acquired IP (i.e., make, use and/or sell the products or services from the acquired IP) without infringing the IP rights of a third party. Freedom-to-operate analysis identifies potential serious legal pitfalls. For example, although a portfolio may have one or more patents related to a certain technology, the claims of those patents may be very narrow, providing limited protection for only a fraction of the desired technological scope. Regardless of the scope of the protection of the patents within the portfolio, whether broad or narrow, a third party may have one or more patents that may prevent the implementation of critically important technology that the buyer believes they are getting. These are known as "blocking patents." Hindrances as to the freedom-to-operate on a patent can seriously impact the value of the patent.

The existence of blocking patents can also impact the freedom-to-operate of a patent. There are two types of patent blocking that can occur: one-way blocking and two-way blocking. In one-way blocking, the dominant patent—which has an earlier U.S. filing date than the second patent—prevents, or blocks, the second patent from being used. Two-way blocking occurs when the patents are mutually exclusive patents existing simultaneously and both effectively block each other, leaving neither one practicable without infringing the other or necessitating licensing the right to practice the other patent. The value of patents blocked by one-way blocking may have little value to prospective buyers if the owner of the earlier patent is unwilling to license the patent, and the blocking patent has a lot of remaining term. However, a prospective buyer who is interested in developing a defensive protection strategy may be very interested in owning blocking patents in order to effectively block their business competitors. Two-way blocking patents may have greatly reduced value unless ownership of both blocking patents can be achieved, or both parties can form a cooperative relationship.

11.1.2.2 Validity of the IP

The evaluation of patent validity involves making an assessment of the novelty and non-obviousness of a patent's claims in view of both the prior art cited during the prosecution of the patent application, or, more importantly, available prior art that was not cited during prosecution. It also involves making a determination as to whether the patent is in compliance with formal requirements of the USPTO, such as the written-description, enablement, and best-mode. Investigation as to whether there has been any inequitable conduct on the part of the inventor(s), assignee(s) or by the prosecuting counsel or agent is also important to evaluating the validity of a patent asset.

11.1.2.3 In Force

Whether the IP assets in a portfolio are in force is important to valuation. For example, the fact that a trademark is not registered should be factored into the valuation of that particular IP asset since the buyer will ultimately be responsible for registering the mark after the portfolio is acquired. It must be understood that both patents and trademarks have additional requirements after issuance. The due diligence must ensure that these additional requirements have been properly addressed to ensure that the IP asset is still in force. For example, a patent with up-to-date maintenance fees will be worth more than a patent that is delinquent in the payment of maintenance fees. Even worse, if the patent's fees were so over due that the patent is deemed to have lapsed by the Patent Office, this requires the payment of a fee to have the patent reinstated and certain rights may have been lost in the interim.

11.1.2.4 Ownership

Clearly identifying the ownership of each piece of IP in the portfolio is paramount to a successful due diligence, as problems revolving around ownership of IP assets can ruin an IP portfolio transaction. Due diligence investigates whether there is a clear chain of title from the inventor, author, or previous owner of the IP as it is important that complete title of the IP passes to the buyer. Confirming ownership can be accomplished by meticulously analyzing assignment and licensing agreements and performing a cross-reference check for any recordation of assignment in the appropriate public records.

11.1.2.5 Scope of Protection

A determination as to the scope of protection will also require an inventory of the IP to be sold or purchased, as well as a record of any encumbrances on the IP. Licenses, assignments or grant back licensing requirements should be identified and declared at the outset of deal negotiations as these things can have an impact on the value of the IP portfolio. Proof that a trademark or copyright in the portfolio has been registered with the Patent and Trademark Office and the Copyright Office respectively will add value to these assets.

Example: Many failures of IP due diligence arise when there is a merger or acquisition between two companies, and the scope of the IP to be acquired by the buyer is not clearly defined or verified by the prospective buyer. In 1998, the Volkswagen AG Corporation purchased the automobile assets of Rolls-Royce Motor Cars Limited. Based on the terms of the sale, Volkswagen purchased the manufacturing equipment and automobile designs from Rolls-Royce. However, after the transaction was completed, Volkswagen discovered that the purchased assets did not include the famous Rolls-Royce® trademark. It was possible for Volkswagen to build the famous Rolls-Royce automobile, but it was legally not able to brand the automobiles with the Rolls-Royce® trademark.

Since IP enforcement is limited to the jurisdiction which issued protection, foreign IP protection in the company's key foreign markets can add value to a portfolio as well. Substantial value can be gained if a prospective buyer has a market in a particular set of countries and the IP sought to be purchased already has protection in those same countries. International protection expands the scope and potential marketability of the IP portfolio, enhancing the value of the portfolio.

11.1.3 Identifying Strategic Value in IP Assets

Performing a comprehensive analysis of the possible sources of strategic value of an IP portfolio can enhance the portfolio's value. The analysis needs to take into consideration both the present and the future corporate strategy (either as a buyer or a seller). There are a number of factors that contribute to the value of an IP portfolio, some of which are extrinsic (or explicit), and some of which are intrinsic (or implicit). A detailed discussion about these two sources of value, particularly as they pertain to patents, can be found in Chap. 12.

Example: Before entering an IP transaction, it is important to verify that the acquired IP will bear the rights the buyer is looking to acquire, especially if the buyer is primarily interested in the particular IP rights because of their unique strategic benefit to the prospective buyer. In 1990, the Clorox Corporation entered into a transaction with the Pine-Sol Corporation to buy Pine-Sol's business and trade identity. Clorox was looking to strategically expand the Pine-Sol brand into other product areas. However, the Pine-Sol® trademark was subject to a preexisting agreement that limited its use to disinfectants only. Clorox learned of this restriction only after the purchase was complete, and while Clorox was able to manufacture products branded with the Pine-Sol[®] trademark, it was unable to accomplish its strategic product expansion goals.

11.1.3.1 Sources of Intrinsic Value

There is much to be gained from various identifiable characteristics of the IP rights for sale. Intrinsic value is value that is inherent in the IP. Some examples of intrinsic sources of value include the quality of a patent (claims, specification, litigation history, etc.), remaining legal and technological lifespan, and quality of the prosecution history. These will be discussed in greater detail in Chap. 12.

11.1.3.2 Sources of Extrinsic Value

Extrinsic value of IP is derived from known, quantifiable sources of value, which are external to the IP; for example present and future cash flows that can be or will be derived from licensing agreements. It is practical value based on what someone is already paying for the IP rights and it is based on what they are willing to pay for those rights in the future. When determining the value of an IP portfolio, a good place to start is with the value of any and all licensing streams generated by the IP in the portfolio. Identifying the amount of the revenue stream that the portfolio is generating per year sets a strong foundation upon which to build acquisition negotiations.

11.1.4 Intrinsic Value of IP-Value Based on Patent Quality

11.1.4.1 Claims

It can be argued that the claims of a patent are the single most important indication of value in a patent because the claims define the scope of protection that the patent will provide upon issuance. Claims need to be meaningful, clear and concise.

Broad Claims vs. Narrow Claims
Broad claims are highly desirable for patents on inventions that are novel and do not face a significant amount of competition through comparable technology. Broad claims cast a wider net of protection and better ensure that the invention, and later improvements to the invention, may be covered by the claim scope. Narrow claims present less risk that prior art will be discovered to invalidate the

claims. Although narrow claims are often easier to design around than broad claims, narrow claims may be very useful in a patent landscape where competitors must have the claimed technology in order to effectively compete.

Independent and Dependent Claims

Independent claims add strength and scope to the patent. They stand on their own and do not depend on any other claim. A single independent claim is the minimum, in terms of claims, required by the USPTO. Dependent claims depend upon an independent claim. A dependent claim incorporates by reference all of the limitations of the claims upon which it depends. If it is determined that a dependent claim has been infringed, necessarily all the claims upon which that dependent claim depends are also infringed.

Claim Clarity

The number of claims a patent has is not necessarily an accurate indicator of patent strength. Rather, the quality of the claims is most important for valuation purposes. Clarity in the claims is important because a patent is meant to put the public "on notice" as to what the protected invention is, with the claims defining the metes and bounds of what is protected. If competitors are unable to clearly identify what the invention is, they may be less inclined to be concerned whether they are infringing on the protected invention.

Word choice is also extremely important in claim construction. The use of precise terms gives the patent claims clarity and eliminates ambiguity regarding the scope of the invention. Terms like "thin", "long", "large", "flexible", "relatively shallow", "suddenly" or "when required" are ambiguous because they are relative terms that lack a precise definition, except if these terms have been given an explicit definition in the specification of the patent.

Example: An illustration of how important it is to draft claims that are meaningful, clear and to-the-point can be found in the chain of events in Finisar v. DirecTV.[1] In 2008, the U.S. Court of Appeals for the Federal Circuit reviewed a dispute on appeal from the Eastern District of Texas where a jury had awarded Finisar a $115 million judgment, finding that DirecTV had willfully infringed Finisar's patent. The district court construed the claim terms "information database" and "downloading into a memory storage device" very broadly. However, on appeal, the Federal Circuit court reversed the construction of the terms, interpreting them more narrowly and found that DirecTV did not infringe the patent. Finisar effectively lost $115 million over eight words that were not precisely defined in the specification.

---

[1] Finisar Corp. v. DirecTV Group, Inc., 523 F.3d 1323 (Fed. Cir. 2008).

11.1.4.2 File Wrapper

Every patent application that is prosecuted before the USPTO develops a file wrapper. A file wrapper is a folder which contains all the papers, correspondences and documents related to the patent application, making up the complete record of the proceedings from the initial application phase to the patent grant. It is the official record of the prosecution proceedings. Every communication between the patentee and the USPTO is recorded and placed in the file wrapper, including office actions, requests for reconsideration, oaths and declarations, information disclosure statements and records relating to examiner interviews. For applications filed before 2003, a copy of the file history of an application is available upon request from the USPTO with the payment of the appropriate fee. For applications filed in 2003 or later, the file wrappers are publically available electronically through the image file wrapper (IFW) system on the USPTO Public Patent Application Information Retrieval (PAIR) website.

The file wrapper of a patent may be an invaluable tool during litigation proceedings. Materials in the file wrapper may be reviewed during a due diligence effort or a litigation for clues as to how prosecution progressed. Counsel typically looks to the file wrapper for material that invalidates the patent, or some sort of admission regarding the scope of the invention, either through the claims or specification. Drafts of a patent application, or uncited prior art in a file make for easy-to-use fodder for an allegation of fraud on the patent office.

As such, it is important to preen a file of unnecessary documents and irrelevant prior art. For instance, a letter exchanged between the inventor and the patent agent or attorney during the application drafting stage in which the inventor expressed that the attorney "got it all wrong and drafted the wrong aspects of the invention," is not a document that should be kept with the patent application documents, nor need it be kept for any length of time.

During an IP acquisition, buyers are looking for patents with clean file wrappers. The more organized a file is—free of anything that is not an official correspondence—the more difficult it will be for opposing counsel to detect anything additional from the materials during litigation. Once a patent becomes involved in litigation, it is illegal to remove any content from the file, as this would be destroying or tampering with evidence.

11.1.4.3 Breadth of Specification

The specification of a patent is the disclosure of the patent. A well-crafted specification will add value to a patent. Specifications that provide thoughtful and thorough definitions for claim terms or provide many embodiments and alternatives of the invention will have more value than a bare-bones specification with few embodiments.

11.1.4.4   Open Continuing Applications

At any time during the pendency of a patent application, an applicant can use the pending application as a parent to one or more continuation applications. Continuation applications are used to expand the scope of protection sought in the parent application. There are three types of continuation applications: continuation applications, continuation-in-part (CIP) applications, and divisional applications. The distinctions between these types of continuations were discussed in Chap. 1 regarding patents.

In terms of identifying value, continuation applications can add value to a patent portfolio. For instance, continuation applications and divisional applications claim the benefit of the parent application's filing date because these types of continuations have the same disclosure as the parent application. The filing date of the parent application, is referred to as the "priority date" or "effective filing date" of the continuation application. A CIP application may also claim the priority date of the parent if the CIP claims are supported by the disclosure of the parent application. If the claims are not fully supported by the parent application, then the CIP application may not be granted the priority date of the parent application because it discloses new matter that was not previously disclosed in the parent application. The new material disclosed in the CIP is essentially a new patentable aspect of the old invention and thus it requires a later filing date; the filing date of the CIP.

There are many reasons why obtaining an earlier priority date on a continuation application is desirable. First, having a filing date that pre-dates prior art is useful during prosecution of the application. Since the filing date is the documented date on which the invention is deemed to have been invented, any prior art reference that is dated after the filing date cannot be used against the application. Secondly, nearly all patent applications are published 18 months after the effective filing date. Publication serves as notice to the public that the invention is being patented. As such, the date of publication for any patent application marks the starting point for when damages may be claimed in the event of infringement.

11.1.4.5   Remaining Term

Consideration should also be given to the remaining life of a patent. This is a two fold analysis: the legal term of the patent; and the technology life-span. IP protection is rendered moot if the technology protected under patent is likely to become obsolete before the expiration of the patent. The value of IP that covers out-dated technology is substantially reduced since no one is inclined to use the technology any longer. Consideration of the technological life-span of a patent would bear more weight in a valuation analysis for patents on technology that rapidly develops but also rapidly becomes obsolete, such as cellular phone technology and virtually any patented computer-related technology. The fact that the technology is obsolete can have a negative impact on the intrinsic value of the portfolio.

For patented technology where the technological life-span is very long, the patent's remaining legal life-span can be a source of value. The term of a patent is presently 20 years from the earliest filing date of the patent. However, in certain circumstances, a patent application may undergo a longer than usual examination by the USPTO. When such delay in the prosecution process occurs, and the delay is due to those actions of the USPTO, a patent term adjustment (PTA) may be granted to the issued patent. When Food and Drug Administration's regulations and procedures stall a patented product from entering the market, a patent term extension (PTE) may be granted to the issued patent by the USPTO. Obtaining a PTA or a PTE can make the legal life of the patent longer than 20 years from the date of filing. PTAs and PTEs are meant to compensate the patent owner for lost time, during which the owner would have had a patent grant had the government not caused a delay. Acquired patents with more remaining years of legal life will provide a longer duration of protection which translates to a greater value for the buyer.

11.1.4.6 Terminal Disclaimers

A basic tenet of patent law is that an inventor is granted a single patent for an invention; not multiple patents. The USPTO may deem a patent application as obvious in light of a granted patent on a similar invention by the same inventor. In essence, the USPTO considers the second patent to be redundant to the first patent and thus not allowable for granting two patents for a single invention. In this case, the USPTO may require the applicant to disclaim a portion of the patent term via a "terminal disclaimer." Usually, the disclaimer will state that the later patent will expire at the same time as the former one. It is also a typical requirement, stated in the disclaimer, that common ownership of the two patents is necessary in order for the patents to be enforceable.

A terminal disclaimer does not affect patent term extension.[2] If the patent has been awarded patent term extension due to delays attributable to the Food and Drug Administration's approval process (for instance with patents on chemical compounds for pharmaceutical drugs), the patent's legal life will end after the expiration of the terminal disclaimer and the patent term extension. Conversely, terminal disclaimers do cut short patent life where a patent term adjustment has been granted to an issued patent.

Regarding the acquisition of patents with terminal disclaimers, there are two important considerations. First, patents with terminal disclaimers hold more value if both are being sold together since the "second" patent is not enforceable if not commonly owned with the first patent. The seller should make sure that they are

---

[2] Merck & Co. v. Hi-Tech Pharmacal Co.482 F.3d 1317 (Fed. Cir. 2007).

being sold together and the buyer should confirm that all patents that have terminal disclaimers are being purchased with their appropriate latter half. The second consideration involves the expiration date of the patents. Since the patents will expire together, the latter patent's active life span will be shorter than the first patent since it was issued later than the first patent.

Example: The pharmaceutical industry provides a great example of how valuable each day of patent protection can be. The joint development effort of Bristol-Myers Squibb and Sanofi-Aventis to develop the prescription blood thinner drug Plavix was a huge revenue generator in the 2000s, generating $9.4 billion in global sales in 2010 alone. Plavix was noted as the second-best selling drug in the world. This was patented in the United States (U.S. Patent No. 4,529,596) and around the world. In the United States, Plavix generated $6,154,000,000 in sales for 2010. That comes to $16.8 million a day!

If the Plavix patent had been subject to a terminal disclaimer, for every day that the patent's legal life had been shortened, it would have amounted to an incredible loss for Bristol-Myers Squibb and Sanofi-Aventis in terms of sales. Once patent protection expires, generic drug manufactures can release competing versions of the drug compound, driving down prices by as much as 70 %.

11.1.4.7 Degree of Potential Apportionment

If multiple patents are being purchased and these patents comprise only portions of a larger technology category, value can be apportioned to the various patents that comprise the technology. For instance, if an IP portfolio is directed at smartphone technology, it can be expected that the portfolio covers a wide range of technology. The portfolio may consist of patents on the smartphone hardware, such as communication systems, transmitters, receivers, antenna, microphone and speaker technologies, as well as other technological aspects incorporated into the device, such as user interfaces and software applications. With so much patented technology contained within a single device, there can be patent infringement of certain patents for specific areas of the technology within the smartphone, without infringement of all of the patents. A competitor may be infringing, for example, a user interface patent, without infringing other patents since the competitor may use entirely different hardware.

Apportionment attributes a portion of the overall value to the specific pieces of IP contained in a product or portfolio. In the event that the patent is infringed, damages can be calculated based on the apportioned value of the patent, aligning the patentee's contribution with the measure of the reasonable royalty. The purpose of apportionment is to eliminate overcompensation to patent holders.

11.1.4.8 Validity

When a patent is issued, it is presumed to be valid, but the presumption of validity is rebuttable.[3] A valid patent is one that is issued by the USPTO after it satisfies the statutory requirements of patentability, (i.e., novelty, nonobviousness, etc.), and no relevant prior art can be found that discloses the claimed invention as being invented by someone else at an earlier point in time. If prior art is discovered that shows that the invention was conceived or disclosed to the public prior to the claimed date of invention (for patents filed before March 16, 2013), or prior to the filing date of the application (for patents filed after March 16, 2013), the prior art can render a patent invalid.

Those IP rights which have been put through the test of litigation, trying their validity yet emerging victorious, have added intrinsic value. For instance, a trademark that is held to be valid by the Trademark Trial and Appeal Board demonstrates the quality and validity of the trademark. Similarly, a patent that survives a validity challenge, or reexamination proceedings, demonstrates the quality of the patent. On the other hand, failed attempts to enforce any IP rights during litigation, or reexamination, can decrease the value of the assets.

Another source of intrinsic value, related to validity, can be derived from how often a patent is cited by others in the relevant field of art. During application examination, the examiner at the USPTO will search for relevant prior art to compare to the invention being examined. If a patent is later cited often as a reference patent during prosecution of later patent applications, it can be an indication of the cited patent's validity and value.

11.1.4.9 Detectability

Detectability refers to how easy it is to make a determination as to whether the invention is being infringed. This includes both detection of infringement and the ease of comprehending what constitutes infringement of the patented claims—such that a jury may understand the infringement. A patent that has clearly identifiable infringers is a patent with value.

Example: Patents for an apparatus or device are claimed in terms of easy to identify elements. Infringement occurs when another device incorporates those same patented elements as the invention. With software programs it can be very tricky or nearly impossible to assess whether infringement by a competitor is occurring without access to the competitor's program source code. For instance, a patented computer program can be copied onto a personal computer, making the infringing use of the software challenging to

---

[3] 35 U.S.C. § 282.

detect, because the use is occurring within a person's private home. Even if the code is not directly copied, which would be copyright infringement, and instead the code is reproduced by copying parts of the code and then adding in work-arounds for the missing code, detection of this infringing activity would also be hard to monitor.

11.1.4.10 Distributed or Divided Infringement

The value of a patent is improved by maximizing the possibilities of direct infringement. Direct infringement occurs when a single actor performs each and every step necessary to practice the invention. However, some modern technology (network applications or Internet transactions) necessarily requires multiple actors, sometimes located in different places, to act in concert to practice the technology. Patent claims on this type of technology, or system, are referred to as "distributed" or "divided" claims; the claims are drafted in such a way that multiple parties must act together to constitute infringement. Recent developments in case law have established that infringement can be achieved by multiple actors acting separately.[4]

11.1.5 Extrinsic Value of IP-Value Based on Issues External to the Patent

Businesses flourish when there is a harmonious relationship between the company's management strategy, R&D strategy, business strategy and IP strategy. That being the case, when a company sets out to purchase an IP portfolio from another company, the buyer is typically looking for IP rights that supplement their own IP and are in alignment with the buyer's business goals. The buyer's business plan may focus on IP acquisition for the purpose of extracting value. For example, the buyer may be acquiring technology and then licensing its use to others. The buyer could also be interested in protecting its market niche by seeking to acquire portfolios for defensive use. The buyer may also be aware of gaps in their own IP protection and may be looking to create a more comprehensive and complete portfolio by acquiring a complementary portfolio.

---

[4] Akamai v. Limelight, 2012 U.S. App. LEXIS 18532 (Fed. Cir. 2012) (en banc).

11.1.5.1 Strategic Value Based on Potential Purchaser's Business Plan

An IP portfolio that is aligned with the strategic business goals of a company has more value to the buyer than if the IP were misaligned. A portfolio that has too large a share of IP related to non-core areas of the buyer's business may be a poor investment. Conversely, an IP portfolio with a high level of synergy with the potential purchaser's business plan can have greater value. Synergy is created by the collection of related IP rights working together. Buyers typically look to create some sort of synergy with their purchase. For instance, unified ownership of a collection of patents that are technologically related may eliminate the hazards associated with blocking patents. For the same reasons, there is value in selling a group of technologically related patents as a bundle, rather than selling the patents individually. Another example of synergy results from grouping IP rights with foreign counterparts and selling them together as a family of IP rights.

Patents that protect basic "building block" aspects of a technology generally have a higher value because they protect fundamental aspects of the research, as well as processing and/or manufacturing techniques upon which down-stream innovation is dependent. Building block patents usually provide a consistent stream of revenue from licensing agreements, because innovation cannot be practiced without practicing the building block patented invention. Businesses benefit greatly from owning the IP rights on building block technologies. As part of their business plan, some companies strategically exploit their building block IP as a way of funding the growth of the business. A company could also use portfolio acquisition as a way of expanding the company into a new area of development.

11.1.5.2 Potential Purchaser's Product Line

Market-driven innovation is important to the growth of business, and the acquisition of a portfolio comprised of IP which is related to the buyer's existing product line, or future product line, can enhance the buyer's business success by allowing it to expand its product lines without infringing the intellectual property rights of others. Expanding product lines is the life-blood of most business models. This type of alignment, between the seller's portfolio and the buyer's needs, can be a huge source of value. A company looking to launch a product may have a need to procure patents that are specifically applicable to the company's product. In this case, the value of that particular patent, or group of patents, is significantly greater to the buyer. By contrast, ad hoc patents with little or no relationship to each other or the buyer's product line may be of little value to a purchaser.

11.1.5.3 Market Value and Market Share

When a company seeks to expand their business operations into a new business or technology area, the acquisition of another company's IP portfolio can have a huge impact on the buyer gaining market share in that new area of expansion. Such a transaction can have greater value to a different buyer who is looking to expand into the new market than to a buyer who is already practicing in that market. In terms of market share, when the buyer looking to expand their business buys a target IP portfolio, not only does he acquire a set of IP rights in the new area (which gives his business a strong foothold to start competing), but he also gets the benefit of assuming the target's market share as his own.

IP acquisition can also have an impact on market value. If an investor considers an acquisition to be strategically advantageous for the business, it may improve the investor's confidence in the company. This may translate into higher stock prices. Similarly, on the flip side of the transaction, disposing of non-core assets in a portfolio, or selling a portion of a business that is not aligned with the primary focus of a business, can improve the market value of the selling company. By eliminating costly portions of the business from the company's operations, the company can improve its cash-flow margins and focus on core areas of business, which enhances investor confidence in the company.

Example: When Google sought to expand its software company into the mobile device market, it gave itself the best possible start it could by purchasing Motorola's mobile device division—Motorola Mobility. The purchase gave Google both access to Motorola Mobility's IP portfolio consisting of 17,000 patents and gave Google Motorola's market share in the smartphone industry, creating a solid place for Google to start competing from. Google was willing to pay $12.5 billion for the opportunity to effectively compete against other smartphone giants such as Apple and Samsung. Google's purchase of Motorola Mobility may have also contributed to an increase in Google's market value. In the two years' time after the acquisition, Google's stock price increased over $150 per share.

11.1.6 Case Study: Astro-Turf's Intangible Assets Sold for Cheap—The Importance of Assessing Intrinsic and Extrinsic Value in Portfolio Valuation[5]

SRI Sports is a company well-known for producing, licensing and marketing AstroTurf, which is a synthetic material used in artificial surfacing applications for

---

[5] *See generally* Weston Anson, AstroTurf: How NOT to Maximize Value, excerpted from The Intangible Assets Handbook (2010) http://www.consor.com/uploadfile/education/pdf/astroturf-case-study_1285842013.pdf

both indoor and outdoor sports and athletic areas. AstroTruf was developed in the 1950s, and popularized when it was used in the Houston Astrodome in 1966, whereby it got its trademarked name ASTROTURF.[6] AstroTurf experienced great commercial success until the early 2000s when a competing product, FieldTurf, came to market and drove AstroTurf out of business. SRI Sports filed for bankruptcy in 2004, and as part of the bankruptcy proceedings, was ordered to liquidate company assets, including its IP and intangible assets, which it planned to sell through an auction.

At the time of the bankruptcy, the AstroTruf IP portfolio consisted of the registered AstroTurf trademarks along with various patents on related technology. The portfolio also included a collection of domestic and international licensing agreements, including an agreement with a Chinese company that generated nearly $200,000 per year in licensing revenue, and a lucrative agreement with the Fédération Internationale de Football Association (FIFA), the governing body for international soccer, to install the AstroTruf on its soccer fields.

Up until 10 days before the scheduled auction, the SRI Sports management team had virtually ignored AstroTurf's IP portfolio and other intangible assets. Troubled that there might be overlooked value in the portfolio, SRI management brought in IP experts to assess whether there was any intrinsic and extrinsic value that could be extracted from the portfolio before the auction. The assessment team found that these assets had not been properly identified nor broken down into logical bundles of IP rights, and that no due diligence had been performed to determine if there were any defects in the rights. With the limited amount of time available before the auction, the team of IP experts was able to verify the licensing agreements and bundled some of the rights together in logical groupings.

Based on the IP experts' analysis, if there had been adequate time to prepare the portfolio to best capture its intrinsic value, it was estimated that the portfolio could have been worth anywhere from $2–5 million, and could have actually realized up to as much as $15 million if fully maximized. However, because the portfolio was not optimized, at auction the AstroTurf IP rights and other intangibles sold for just less than half a million dollars—barely twice the extrinsic value of the Chinese license which was single-handedly generating $200,000 in licensing revenue per year.

11.1.7 Consider the Risks Associated with IP Portfolio Acquisitions

There can be many different types of risk involved in the acquisition of an IP portfolio. A seller should identify and attempt to mitigate any risks associated with the contents of the portfolio, but when that is not possible, the seller should factor

---

[6] U.S. Registration No. 1,930,544. Owner: Southwest Recreation Industries, Inc.

the risks into the valuation of the portfolio. A buyer expects a reduction in price if they are purchasing IP that requires further development in order to achieve a final product. It is possible that the technology behind the IP may flop once developed or may not be scalable for commercialization purposes. Similarly, the buyer may not be able to gauge future market success of the technology once it is developed and commercializable. Less developed intangible assets have higher risk associated with them. The buyer considers questions like:

- How will the market perceive the final product?
- Will customers accept the technology?
- Will the product be affordable once the costs of developing the technology are factored into the purchase price?
- Is this a technology which requires regulation by a government agency?

These considerations may have an impact on the value of the IP behind the technology. A buyer does not want to purchase a portfolio that has a lot of grounds for potential liability. The less liability associated with the purchase, the more it is worth to the buyer. A seller should take steps to reduce the amount of liability in their portfolio. Performing due diligence to verify the ownership rights of the IP, procuring validity opinions from independent IP counsel, and acquiring IP insurance can all help reduce the amount of liability in a portfolio and improve buyer confidence.

Chapter 12
Valuing Technology Companies

Simply stated, the value of a business or asset is represented by the future economic benefits that will inure to a buyer. The value of a business depends on an estimate of the future economic benefits, the period of time for these benefits and the rate of return required by the buyer based on the industry and company specific risks.

Business valuation is a process used to estimate the economic value of an equity interest in a business. There are various elements of a business valuation which includes the definition of value to be used, economic conditions, financial analysis, normalization of earnings, valuation methods and approaches, industry and company risk factors and type of interest owned.

Valuations of traditional, mature companies can be challenging in spite of using industry accepted valuation methodologies; however, technology company valuations take this challenge to a new level. Traditional companies tend to have different characteristics than technology companies. Traditional companies have

(i) more hard assets, such as property, plant and equipment,
(ii) more historical financial data,
(iii) less uncertainty in their projections and forecasts, and
(iv) less volatility in their cash flows.

These characteristics can have a significant impact on the way companies are valued. Technology companies can range from early stage to commercialization to growth or expansion stage to mature stage of a company's life cycle. The information available tends to be rather limited and there tends to be significant uncertainty with regard to both the product and any financial data.

By: Robert W. Fesnak, CPA, ABV, CMA, CVA, Partner Fesnak and Associates, LLP

G. B. Halt, Jr. et al., *Intellectual Property in Consumer Electronics,*
Software and Technology Startups, DOI: 10.1007/978-1-4614-7912-3_12,
© Springer Science+Business Media New York 2014

12.1 Purpose of a Valuation

A business valuation may be needed for several reasons. The most common purposes are:

- Gift or estate tax purposes
- Determining the sale or purchase price of a business
- Allocate purchase price among acquired assets
- Buy-sell agreement purposes
- Stock option and other share-based compensation plans
- Obtaining capital
- Damage calculations in a litigation.

Technology company valuations are required for the following situations:

- Equity ownership must be fairly allocated at each equity round. This requires an overall enterprise valuation as well as a valuation of the specific classes of equity such as preferred and common interests.
- Technology companies usually grant stock options, restricted stock or similar securities in order to attract and retain the best talent. The underlying enterprise value and the value of the class of security to which the incentive relates must be determined in order to value these incentives. There are both financial and tax implications of such incentives.
- In order to obtain subordinated or mezzanine debt financing for a technology company, debt holders generally require warrants as an additional incentive to offset some of the risk they are taking. The value of such warrants must be valued for financial reporting.
- The Financial Accounting Standards Board has issued accounting pronouncements that require the fair value of stock options, warrants, restricted stock or similar securities to be recorded in the company's financial statements.
- The Internal Revenue Service has issued regulations (IRC 409A) that require companies to grant options at an exercise price equal to or greater than the underlying security value on the date of grant or suffer adverse tax consequences. In order to determine whether granted options meet this requirement, the fair market value of the security subject to the option must be valued.

There are different definitions of value that can be used. The valuation results can lead to different values based on the definition. The definitions include fair market value, fair value, investment value, strategic value and liquidation value. Our discussion in this section will focus primarily on fair market value.

12.2 Methodologies

The valuation of businesses utilizes several methodologies which include cost, income and market approaches. The cost or asset based approach values a business based on the difference between the fair market value of the business assets and the fair market value of its liabilities, which represents the fair market value of the Company's equity. In essence, the cost to replicate or reproduce it. The theory being that a buyer would only pay an amount not to exceed the cost to replicate the business (i.e. functionality of technology; management; market share; reputation; etc.). The income approach values a business based on its expected earnings or cash flows. The market approach determines value based on an analysis of sales and investments in comparable businesses. Comparable businesses must be similar with regard to industry, size, capital structure, management and position in the market.

Factors used to determine the value of technology companies have changed over the years. Valuations of technology companies continue to utilize the income and market approaches; however, their valuations are focused more on near-term breakeven or profitability, strong management, experienced board of directors, size of market, strategic "partnerships" and a technology or service with a sustainable competitive advantage. Strategic partnerships and a technology with a sustainable competitive advantage will have a significant impact on the value of a business. Management and their advisors must develop strategies to generate these sustainable competitive advantages which will allow them to dominate markets. This is generally done by analyzing competition, positioning the company in a unique market space, performing market research and implementing appropriate strategies. The proper execution of these strategies is the critical component.

12.2.1 Cost or Asset-Based Approach

The cost approach is also known as the asset-based approach. This approach is not a reliable indicator of value for technology type companies. Historical costs do not necessarily represent the value of the company or even the value of the asset for which the costs were incurred.

In the asset-based approach, all of the subject company's assets and liabilities are analyzed and valued separately. The company's assets include tangible and identifiable intangible assets. However, it does not include a value for goodwill. Goodwill (GW) is the residual value of the enterprise value less tangible and identifiable intangible assets.

Most companies can maximize their value based on their earnings level and not just their net asset value. A technology company's value will consist primarily of its patents, in-process research and development, customer list, data base, trade secrets, other intellectual property value and goodwill. A market or risk adjusted income approach is usually more appropriate for technology companies.

12.2.2 Income Approach

The income approach can be utilized by (i) capitalizing the historical earnings of a company or (ii) discounting future earnings or cash flows ("DCF") of a company. The DCF approach is based on the premise that the total enterprise value of an entity is the present value of its forecasted future net cash flows, plus the present value of a terminal or residual value. The terminal value is the value of the future cash flows at the end of the discrete forecasted period. Business enterprise value is also referred to as the market value of invested capital ("MVIC"). MVIC includes the value of both interest bearing debt and equity or total invested capital. Equity value is the MVIC less the value of interest bearing debt. This method requires that a stream of earnings or cash flows be reliably forecasted into the future and that a terminal value assumption be made. The amounts of forecasted cash flows and the terminal value are then discounted to the valuation date using an appropriate discount rate.

The historical earnings for technology companies are usually not representative of future earnings. Technology companies tend to incur significant costs early on and then, after product commercialization, grow rapidly. This would mean that the DCF would be more appropriate to value the Company vs. the capitalization of historical earnings method.

The discount rate used to determine an entity's MVIC is based on the weighted average cost of capital ("WACC"). The discount rate reflects percentages of total debt and equity to capital and the rate of return that an investor would require based on the risk associated with the investment. The WACC includes both the cost of debt, after tax, and the cost of equity.

12.2.3 Market Approach

There are generally two primary market methods, the guideline public company method and the guideline company transaction method (also known as guideline merged and acquired company method).

A guideline public company method analysis is intended to derive valuation multiples from traded prices of public companies that are relatively comparable to the subject company. The value of the subject company is then estimated by applying an appropriate valuation multiple derived from the public companies to the appropriate benefit streams of the subject company. These benefit streams could be revenues, pre-tax income, operating income or earnings before interest, taxes, depreciation and amortization ("EBITDA"). A valuation multiple is usually a multiple computed by dividing the MVIC and/or the equity value by the relevant economic factor.

The value derived by the guideline public company method is on a marketable, minority interest basis. This means that the value needs to be converted to a non-marketable value for privately held companies and a controlling interest value for an enterprise value. An example of this is as follows:

Marketable, minority interest basis	$100
Control premium of 20 %	20
Marketable, controlling interest basis	120
Discount for lack of marketability of 15 %	(18)
Non-marketable, controlling interest basis	$102

The guideline company transaction method is another market approach. This method is intended to derive valuation multiples from the acquisition prices of public or private companies that are relatively comparable to the subject company. The value of the subject company is then estimated by applying an appropriate valuation multiple derived from the transactions to the appropriate benefit streams of the subject company. Factors to consider with regard to the comparability of market transactions to the subject company are:

- Terms of the selected transactions
- Consulting agreements vs. purchase price
- Number of companies in the population
- Comparability of data—for example, similar growth trends, profitability
- Stock vs. asset transactions
- Dates of the transactions in the population

The transaction prices may represent fair market value, investment value, strategic value or a mix of these. Understanding these guideline transactions will allow you to better determine which definition of value they represent and thus you can apply the data appropriately to the subject company.

12.3 Normalization of Earnings and Cash Flows

The concept of normalization adjustments is to present the financial data such that it reflects the expected future economic benefits to an investor or a buyer. This means the elimination of non-recurring revenues and expenses from the subject company's financial statements. Normalization includes adjustments for such items as excess officer compensation (compensation that exceeds industry norm), loss of a major customer or sales contract, one-time expenses, unusual research and development costs, etc. It is critical to use the "true" expected cash flows of the subject company for valuation purposes. No buyer ever bought a company for its past cash flows. They are buying the future cash flows. Historical cash flows may be used as a proxy for future cash flows; however, the data needs to be normalized. If historical cash flows are not representative of the future then forecasted results will be used to value the company.

12.4 Impact of Risk on Value and Cost of Capital

Risk has a major impact on value. Risks that need to be considered include external risks such as the economy and the industry of the subject company and internal risks that are specific to the subject company.

Risk is generally defined as the degree of uncertainty as to the realization of expected future benefit streams such as earnings or cash flows. Risk is directly correlated to the amount of uncertainty and volatility of the expected economic benefits. The higher the uncertainty, the higher the risk. The higher the risk, the higher the required rate of return an investor will want and thus the lower the multiple of earnings he or she will pay. The present value of the cash flow and thus the company's value decreases as risk increases. This creates an opportunity for the subject company to enhance its value over time even if its revenues and earnings are flat by reducing its risk profile. This can be done by focusing on such items as:

- Stability of earnings
- Size of company
- Protection and leverage of intellectual property
- Depth of management; dependency on key people
- Stability and skills of workforce
- Level of differentiation
- Competition (barriers to entry, switching costs, etc.)
- Diversification of products, services, geography
- Dependency on key customers and vendors
- Litigation.

Cost of capital is composed of the cost of equity plus the after-tax cost of debt. The discount rate is derived from the cost of capital and is used in the DCF method.

Risk is addressed in the valuation process through the discount or capitalization rates that are applied to the earnings or cash flows. The cost of equity consists of the risk-free rate, an equity risk premium, an industry risk premium, a size premium and a company specific risk. The company specific risk allows the valuator to customize the cost of equity for risks specific to the subject company where these risks are not already included in the industry risk premium.

12.5 The Uniqueness of Early Stage Technology Company Valuations

To fully understand valuations of early stage technology companies, knowledge of the different economic and even social environments in which these companies operate is useful. In addition, since most technology companies are funded by venture capital firms, knowledge of the venture capital industry can be beneficial.

Due to the lack of historical financial data and the high level of uncertainty, the reliance on traditional methods for valuing early stage companies is not useful. In the late 1990s and 2000 it was common to value early stage technology companies on non-financial metrics. Some of these included:

- Price to page viewers
- Price to subscriber
- Price to downloads
- Price to future revenue
- Price to full time employees.

These were not reliable benchmarks for value.

Determining the value of an early stage technology company requires the use of alternative approaches vs. a company that has revenues and is profitable. The approaches require a greater understanding of the qualitative aspects of the company such as the management team, business model and size of the specific market. Early stage technology companies tend to (i) lack an operating and financial history, (ii) their product or market is not yet validated and (iii) value tends to be primarily in technology/patents/intellectual property, market size and ease of access to capital markets. Technology companies tend to require rapid growth or risk possible loss of an opportunity. The primary risks associated with technology companies are:

- Will the technology work as expected?
- Is there a demand in the marketplace for the technology?
- Will the marketplace accept the pricing strategy?
- Is management capable of executing the business plan and achieving corporate goals?

The more information on these factors the less uncertainty. Less uncertainty, like any other investment, means less volatility and a better ability to substantiate value. Obviously having trailing revenue answers the first three of these questions and provides financial data to form more reasonable and reliable forecasts. However, early stage companies generally only have a beta version of their product. This is why other types of metrics and qualitative data must be utilized to determine a company's value.

Since the cost, income or market valuation approaches are usually not appropriate for early stage technology companies where the business is in a pre-revenue and thus pre-earnings stage, more non-traditional methods need to be used to determine value. These include:

- Stage of company—there is usually a range of value based on the stage of the company as to its product development and time to revenues. Data is generally available as to the valuations of other early stage companies at a similar stage.
- Average dilution for early stage investment round—Early stage companies typically give up 25–40 % of the company in a seed stage round of funding. The post-money valuation can be estimated by dividing the amount of funds

generated in the round by the dilution percentage. The pre-money valuation is then calculated by subtracting the new funding from the post-money valuation. So if a company receives $1 million in new funding and gave up 30 % of the company then the post-money value would be $3.333 million ($1.0 m/0.3). Its Pre-money valuation would be $2.333 million ($3.333 m–$1.0 m).

- Prior round financing—A prior round value may be able to be utilized to estimate the value of a technology company if no significant events have occurred since the last funding transaction.

Technology company valuation approaches and methods are summarized in the following table:

Table 12.1 Technology Company Valuation Approaches and Methods

Stage of development	Typical type of financing	Typical valuation approach	Valuation methods
1. Pre-revenue; no product; initial product research and development (more research than development); still building management team	Seed capital (common stock) from friends and family, angel groups, family offices	Market	Option pricing model ("OPM"); based on stage of company
2. Pre-revenue; substantive cost data; product development in mid-stream;	Series A round (preferred stock); usually angel groups or early stage venture capital ("VC") funds	Market	OPM; stage of company; prior round
3. Key development milestones are met; beta model is complete; market well understood; still pre-revenue but may show customer interest	Series B and/or C rounds (preferred stock);	Market; income	OPM; prior round; discounted cash flow ("DCF")
4. Product revenue; operating at a loss	Series C and/or D rounds (preferred stock or convertible debentures)	Market; income	OPM; prior round; DCF
5. Generating significant revenue; operating at a profit or breakeven	Additional series of preferred stock	Market; income	Guideline public company data; guideline market transactions; DCF
6. Financial history of profits and positive cash flows	Liquidity event or IPO; mezzanine financing	Market; income	Guideline public company data; guideline market transactions; DCF

12.6 Allocation of Enterprise Value to Senior and Junior Equity Interests

To this point, we have referred to the value of the overall enterprise. Technology companies, however, tend to have complex capital structures as the result of their capital-raising activities needed to fund long periods of time until products are developed and introduced into the marketplace.

Capital is often raised by issuing common stock, preferred stock, convertible debt, options and warrants. In certain cases, the value of one class of securities must be determined. For example, the per share value of common stock must be determined for purposes of granting options for common stock. It would be incorrect to divide the overall enterprise value by all shares outstanding to arrive at the per share value of the common stock due to the preferences of senior securities and the dilutive effect of options and warrants. Preferences of senior securities could include items such as dividend and liquidation preferences, voting rights or other control features, conversion features, participation features and redemption rights. The exercise of options and warrants could create additional shares outstanding, thus reducing the per share value of the common. Three complex, yet common, valuation methodologies are used to allocate value among the various classes of capital giving consideration to the preferential rights of senior securities.

12.6.1 The Current Value Method

Assumes senior security holders would monetize their value through their liquidation preferences and participation rights in an assumed imminent liquidity event. The remaining enterprise value would represent the common stock value. This method is only used where there is an imminent liquidity event.

12.6.2 The Option Pricing Method

This method relies on financial option theory to allocate value among different classes of stock based on a future "claim" on value. This method is generally used for early stage technology companies where future expected earnings are not reliable.

12.6.3 The Probability-Weighted Expected Return Method

Share value is based on the probability-weighted present value of expected outcomes such as a future liquidation event, IPO or continued operation as a private enterprise as well as the rights of each class of securities. This method is used for mature technology companies with reliable future expected earnings.

12.7 Leverage and Monetization of Intellectual Property

Intangible assets, which includes intellectual property, continues to represent a larger portion of the value of companies in general, let alone technology companies. Intangible assets represent approximately 80+ % of the market value of the Fortune 500 companies.

An intangible asset will have the following attributes:

- It is subject to specific identification.
- It should be subject to legal existence and protection.
- It should be subject to the right of private ownership and be legally transferable.
- There should be tangible evidence (contract, source code, etc.).

As mentioned earlier, intangible asset values generally represent a significant portion of the value of the company. These intangible assets include such items as:

- Marketing related—trademarks, trade names, brand names
- Technology related—process patents, product patents, technical documentation, technical know-how
- Artistic related—copyrights, literary works
- Data processing related—proprietary computer software, software copyrights, databases
- Engineering related—product patents, trade secrets, proprietary documentation
- Customer related—customer lists, customer contact information
- Contract related—favorable supplier contracts, license agreements, non-compete agreements
- Human capital related—employment agreements, trained and assembled workforce
- Location related—leasehold interests, easements
- Goodwill related—enterprise GW, personal GW, excess of enterprise value over tangible and identifiable intangible assets.

Intellectual Property (IP) is a specialized type of intangible asset. IP is commonly defined as intangible assets that are either creative (e.g., trademarks, copyrights, computer software, etc.) or innovative (e.g., patents, trade secrets, etc.).

In today's market, the ability to understand the value drivers of intellectual property is critical. This understanding will impact not only the company's IP strategy but also how best to monetize the IP. There is a clear relationship between the quality of a company's IP, how well it is protected and a company's value. IP rights provide an opportunity and potential to generate revenues and profits. It does not guarantee revenues or profits. Execution of an IP strategy, taking advantage of these IP rights, allows the technology company to create value and monetize such assets.

12.7.1 Value of IP and Impact on Technology Companies

The value of identifiable intangible assets, and thus technology companies, is influenced by such factors as (i) market share, (ii) market potential, (iii) economic life of the IP, (iv) cash flow from the IP, (v) life-cycle stage, (vi) market position and (vii) barriers to entry. By focusing on these and possibly other factors that influence value the intangible assets of a technology company can become more valuable. An example of this is where a company is able to extend the economic life of a patent or trade secret through threatened enforcement actions or significant barriers to entry.

The valuation approaches for intangible assets are similar to that of an enterprise in that they include cost, income and market. However, the methods within each category can be different.

- Cost approach includes

    - Reproduction method—estimated cost to reproduce an exact replica of the subject intangible asset using same design, standards and materials.
    - Replacement method—estimated cost to replace with an asset that has the same functionality as the intangible asset.

Replacement is most likely more relevant for IP than the reproduction method. A seller's cost of IP is irrelevant to a buyer. The only cost that matters is the buyer's "substitution cost," i.e., the cost to replicate the functionality of the technology or IP. A buyer will not pay more than the cost to develop a substitute IP.

- Market approach includes

    - Sales transaction method—estimate value based on actual market transactions/sales of comparable or guideline intangible assets in arms-length transactions.
    - Relief from royalty method—estimate value of the subject intangible asset based on the cost saved by owning the intangible asset vs. having to pay a market royalty rate for the use of it. The present value of these savings are then capitalized to determine value. In using this method, arms-length royalty

rates are analyzed. The licensing transactions selected to determine the royalty rate should be for assets with similar risk and investment return profiles.

- Income approach includes

  - Discounted cash flow method—estimate value based on the present value of the expected cash flow associated with the ownership and use of the subject intangible asset. Since the cash flow is derived from the use of not only the subject asset but also all other assets of the company, an economic charge must be deducted from the future cash flows to account for the utilization of the other company assets.
  - Comparative income differential method—estimate value by comparing the income generated in two scenarios; one with and one without the use of the intangible asset being valued. The present value of the differential in income equates to the intangible asset value.
  - Monte Carlo and Real Options methods—Both methods rely on the DCF method. The Monte Carlo method estimates key variables within certain ranges of outcomes and with related probabilities. The Real Options method evaluates risk and returns in increments.

12.7.2 IP and the Enforcement of Rights

IP can be protected numerous ways from (i) well written, defensible patents to (ii) non-disclosure agreements and other protective measures for trade secrets to (iii) non-solicitation agreements for customer lists and on and on.

Historically, only large corporations could afford to wage the frequently long-term, expensive patent litigation wars. Smaller firms often capitulated as high costs to litigate meant they would be forced to rely on scarce cash reserves and available debt.

IP insurance has been around for some time but it has largely been defensive in nature. It covers the cost of legal fees and damages (up to policy limits) associated with responding to allegations of IP infringement from a third party.

But what about the smaller or medium sized firm who develops an innovative technology? How can they protect themselves from an infringer who is significantly larger than them?

The value of IP is dependent on the owner's ability to enforce the rights imbedded in the IP, whether that is a patent, trademark, confidentiality agreement or non-compete agreement. Without liquidity and the capital resources to enforce these IP rights the value of the IP can be diminished. One way to enforce IP rights is through enforcement insurance; it is designed to reimburse the insured for legal expenses incurred (up to a specified amount) to enforce its IP rights. The underwriting process, in theory, provides additional validation of the IP and the insurance proceeds provides the means and liquidity for the company to enforce its rights.

The insurance may actually discourage potential infringement by demonstrating financial wherewithal and could reduce the pressure to settle due to otherwise limited resources. Additional benefits could include:

- Reduces risk and uncertainty
- Enhances the value of IP
- Prevents unexpected use of cash resources
- Provides additional protection to the primary assets for investors
- Accelerates commercial development and licensing opportunities since IP is protected

As the value of IP increases so too does the associated threat of infringement.

12.7.3 IP Monetization

Monetization is the process of extracting value through a liquidity event. The monetization of IP is the process of extracting value from the business' IP. This usually entails the sale of the IP or the entire business. However, there are other ways to monetize IP.

An interesting new development is the use of the IP as collateral for loans. The author was recently engaged by a financial institution to value brands and trade names being acquired by a company. The primary assets of the new venture were various forms of IP as they outsourced the manufacturing and other operational aspects of the business. The brands had strong cash flows and the financial institution loaned funds for the acquisition based on these cash flows and the value of the IP.

In addition, there has been various discussions with regard to private investor groups that are designed to provide loan guarantees to growth companies that own significant intangible assets.

Certain insurance and financial companies have provided credit enhancements to banks and other investors for royalty and non-royalty generating patents, trademarks and copyrights, which were then used as collateral in a structured financing. These companies will typically evaluate the strength and breadth of the IP, value the IP and issue a guarantee to the financial institution or investor group. They will then charge a fee for the guarantee/credit enhancement. It has generally been used for larger more mature IP but the author is seeing an interest in utilizing this concept for smaller companies as well.

The utilization of (i) lending on IP, (ii) having investors use intangible assets as collateral for a loan guarantee or (iii) utilizing a credit enhancement insurance instrument are just some ways to generate capital resources and monetize these assets. It also provides less expensive cost of capital through use of senior or mezzanine debt along with equity capital.

The author believes as companies become more virtual (outsourcing more and more functions) and focus mainly on building value in their technology and IP, the use of asset based IP lending and/or investor guaranteed debt should increase. It behooves technology companies to focus on what creates value and this is usually IP and intangible assets in general, not necessarily hard assets.

The goal of the ideas discussed in this section is to provide insight into ways to protect, optimize and monetize the real value of a technology company, its intangible assets. However, there is no guarantee that even if these strategies were used, that financing would be available.

12.8 Summary

Although traditional valuation approaches are sometimes used to value technology companies, special techniques are required to compensate for the unique characteristics of these companies.

Due to the nature of the product development and commercialization cycles, and the lack of revenue and earnings metrics for guideline market comparisons, alternative approaches are sometimes necessary. Financial forecasting is difficult and risk rates are high for the application of discounted cash flow models. Probability-adjusted or risk-adjusted forecasts present a more refined analysis of expected future events and thus a more refined estimate of enterprise value. In addition, capital structures are complex and consideration must be given to the preferences of senior securities and dilutive effects of options and warrants when determining the value of common stock.

In addition, there appears to be a growing interest in finding ways to better monetize a company's intellectual property.

Chapter 13
Financing and IP: Equity Financing, Debt Financing, Collateralization and Securitization of Intellectual Property Rights

From the moment of inception of a great idea, there is an inclination to protect and nurture that idea, striving to turn it into a commercial success. At the start of an entrepreneurial adventure, IP rights are procured on the idea and a business is formed. As the new company moves from the design and conceptualization stages, through product development and finally to market, there are many stages of growth which all require financing in some capacity. As the idea moves from the seed stage of development, through the growth and expansion phases and ultimately to maturity, different financing options become available to the IP owner.

There are a multitude of ways to monetize IP assets and different strategies need to be employed based on the size and capital structure of the company. Large, well-established corporations, brand-new startup companies, and universities and research centers have different needs and constraints on their ability to monetize. Regardless of the situation, leveraging an IP portfolio can provide many opportunities for monetization of those assets.

13.1 Benefits of Leveraging IP

Leveraging is a means of extracting value from an IP portfolio in a way that is to the competitive advantage of the IP holder. There are many ways to leverage value from an IP portfolio. Methods for extracting cash from a portfolio that immediately spring to mind include exploitation of the IP rights, licensing, or the sale of the portfolio. However, other avenues of monetization are available to an IP portfolio holder which may not be readily apparent. For example, an IP portfolio could be used as a source of equity, as a debt instrument, as collateral or as a security interest; any of which can be leveraged in various financing options.

G. B. Halt, Jr. et al., *Intellectual Property in Consumer Electronics,*
Software and Technology Startups, DOI: 10.1007/978-1-4614-7912-3_13,
© Springer Science+Business Media New York 2014

13.2 Equity Financing

Due to the risky nature of lending money to a startup, banks will typically hold off on providing financing to a small business venture until it has successfully entered the growth phase of development. In the interim, smaller businesses and startup companies need alternative financing sources. Equity financing is a good option for raising startup capital. Venture capitalists, angel investors and private equity partners are important sources of financing for small ventures. These types of financiers are often individuals, but can also be institutional investors, that are interested in investing in small entities that have great potential for growth. These investors, as part of their financing negotiations, often bargain for a stake in the company's success by providing the company with up-front capital in exchange for an ownership interest in the company, i.e., common or preferred stocks issued to the financier. One of the benefits of equity financing is that the business raises funds without incurring debt and the business is under no obligation to repay specific sums to the investors at specific intervals of time. One of the main disadvantages to this type of financing is that it dilutes the equity of the company.

13.2.1 Venture Capitalists

Venture capital (VC) is a form of funding for new business ventures that is often conditional—meaning the funds usually come with strings attached. A few examples include instances where VC funds are granted in exchange for an interest stake in the company, a seat on the board of directors, the right to approve loans on behalf of the business, authority over the hiring/firing process, or various promises that the VC will be included in all major business decisions. VC funds are typically derived from a pool of professionally managed funds contributed by either individual venture capitalists or institutional investors. When VC funds are administered to a business, they typical range between $500,000 and $10 million.

Venture capitalists often have certain criteria that must be met or demonstrated by the business venture before the VCs will consider it a viable and worthwhile investment. For instance, VCs are typically looking to invest in companies with potentially large and lucrative markets since the VCs' primary interest is to ultimately obtain a high return on its investment. VC funding is often limited to established technologies that are beyond the proof of concept stage; VCs are more inclined to invest in entities that are at the product development stage or at the production and marketing stages of commercialization since the technology is well-developed by that point. Also, there are many VCs that prefer to invest only in companies that have an IP portfolio.

13.2.2 Angel Investors

There are some affluent individuals, often referred to as angel investors, who are interested in investing privately in a small business during the seed or early startup stages of the company's growth—the stages of a business' development which are least certain and have the highest risk. Because the risk is so high, angel investors often require a return on their investment at a higher rate than VCs. It is common for angels to invest in industries that they have past experience in, and will generally invest in businesses that are local to the angel since the angel often relies on his or her established network of professional service providers, such as accountants/tax specialists, legal counsel and lenders, when assisting the investment venture's growth.

Like VCs, angels provide funding to companies in exchange for convertible debt or ownership equity, and contribute anywhere from $100,000 to $500,000 in private equity funds. Angels prefer businesses with a working prototype of the product, although some are persuaded to invest based on a proof of concept. Angels often look for at least the beginning steps towards IP protection of the product, such as pending or provisional patent applications, when deciding to make an investment.

13.2.3 Private Equity Partners

Private equity firms are investment sponsors that invest in companies in the early stages of development with IP assets integral to the company's intended market. Like angels, the private equity (PE) investors typically invest in industries with which they are familiar. PE partners make investments between $250,000 and $25 million. Private equity investors work with the company to create a lucrative exit strategy for the PE investor where both the business and the PE investor mutually benefit. Exit strategies commonly employed by PE partners include leveraged buyout by another investor or a merger or acquisition of the company. Commencing an initial public offering of the company is also a common exit strategy for PE investors.

Example: Altitude Capital is a boutique private equity firm that invests in companies with IP portfolios consisting of valuable intangible assets and IP. The motivation behind such investment is to create a competitive advantage by creating value in the company based on its IP rights. Intrinsity, Inc., one of Altitude's portfolio companies, developed a key high-performance microprocessor technology—its proprietary FastCores technology—which Altitude invested $11 million into, in exchange for shares of preferred stock in Intrinsity and secured notes with warrants.

13.2.4 Initial Public Offerings (IPO)

Once a company is at the point of being marketable, it can initiate an IPO, with approval of the board of directors, to raise additional capital. An IPO is the first public sale of stock by a company, thus moving the company from a private business entity to one that is available for public trading. A company can use its IPO earnings to pay off past debts, reinvest in the company or pay out a dividend to its new shareholders. Being a publically traded company can have its advantages. For example, the company can raise capital via the sales of its securities through primary or secondary markets, where privately held companies lack the benefit of free trade. Public trading activity can cause share prices to go up, creating investor confidence in your company. Also, publically traded companies tend to be in the news more often than private companies, which translates to free publicity for the company.

Being a public company also has disadvantages, such as the public disclosure of financial statements due to the mandatory filing requirements of the Securities and Exchange Commission (SEC). Publically traded companies are required to submit to the SEC an annual Form 10-K, which contains detailed information on the company's financial performance, as well as other disclosures, such as shareholder patterns, quarterly and annual financial statements and profiles of the directors of the company. Additionally, there are many expenses associated with going public, all of which can tally up to a significant sum. Furthermore, if an IPO is not successful, the company can lose money.

13.2.5 Case Study: Facebook's IPO

To celebrate its eighth birthday as a company, Facebook's CEO and founder Mark Zuckerburg decided to file paperwork with the Securities and Exchange Commission for Facebook to participate in an initial public offering. The filings were made on February 1, 2012, setting an IPO date for some time in late May of the same year. A flurry of pre-IPO excitement abounded as the Facebook IPO date approached. The hype was largely due to the high valuation estimates of Facebook initial stock offering price, and the recent successes of other Internet company IPOs, like Google and LinkedIn. Investment analysts were predicting that the Facebook IPO would be the "the next Google" and that Facebook would easily see the price of its shares double within the first day just like LinkedIn during its IPO. Thirty-three banks and underwriters participated in the determination of the Facebook offer price per share, with investment banker giants Morgan Stanley, J. P. Morgan, and Goldman Sach leading the charge. The social networking site was one of the largest Internet IPOs in history, with the initial stock price set at $38 per share, this stock price resulted in a company valuation of $104 billion—the largest valuation in history for a company going public.

On May 18, 2012, Facebook, Inc. entered the publicly traded sphere. The company's shares were set to begin trading on the Nasdaq Stock Market around 11 a.m. ET under the symbol FB. However, Nasdaq encountered numerous glitches in the first few hours of trading, beginning with a half an hour delay in the commencement of Facebook trading. At 11:30 am trading finally started, with more than 80 million shares changing hands in the first 30 seconds. Shortly thereafter, traders noticed delays in their orders going through, and some traders found that they were getting shares, but at a higher price than they expected to pay. Some orders were never placed in the Nasdaq system due to the technical errors that Nasdaq claimed maimed the trades. Confirmations of successful trades were not issued immediately, leaving traders unsure as to how many Facebook shares they had actually bought or sold. Many traders found that they had taken losses due to Nasdaq's error and sought from Nasdaq a plan as to how Nasdaq would compensate them.

At closing bell, shares were valued at $38.23, only $0.23 above the IPO price, and it was quickly decided throughout the investment community that Facebook had been overvalued. The stock lost over a quarter of its value within a month and dropped to less than half its IPO value within 3 months of the IPO. Not only were there problems with Nasdaq's trading system, but there were also concerns that Facebook itself had wrongly engaged in some inappropriate pre-IPO activities. It was alleged that sensitive information about Facebook's financial outlook was selectively disclosed to several large banks prior to the IPO. Facebook was also accused in undertaking a "pump and dump" scheme—a form of securities fraud. In a "pump and dump" scenario, the "pump" occurs when an IPO company rises the price of its initial shares by means of insider manipulation and press releases that sound too good to be true. The "dump" follows when insiders sell their shares after the stock has sucked up every last cent worth of value, leaving ordinary investors holding stock that is worth considerably less than they were anticipating if all had been fair and transparent.

Facebook's IPO has been labeled as an "over-hyped disaster." Many of the players involved in the Facebook IPO have found themselves involved in litigation as a result, which at the request of Facebook, has been centralized to be heard by a single U.S. District Judge in New York. U.S. District Judge Robert Sweet has been placed in charge of hearing all cases related to the botched Facebook IPO, including cases against Nasdaq regarding the problems experienced during the initial trades. By October of 2012, there were 33 pending suits related to Facebook's IPO. Facebook stands to lose a lot of money as it defends itself from the claims of unhappy investors.

13.3 Debt Financing

Debt financing is a more attractive means of financing for larger business entities. In debt financing, money is borrowed with the intent that is it to be repaid over a fixed period of time, with interest. Interest paid on the loan is tax deductible for the borrower. Debt financing is referred to as short-term and long-term debt financing. Short-term will be repaid in full within one year of borrowing, and long-term will be repaid in full more than one year after borrowing. Debt financing is essentially a loan whereby the lender retains no ownership interest in the business in exchange for the loan. Instead, the lender earns interest on the amount borrowed. When the borrower is a small business, it is common for the lender to require a personal guarantee based on the business owner's credit history as an individual when making the loan to protect its investment.

A creative financial solution is IP-based venture debt financing. IP-based venture debt financing is obtained by venture-backed entities that is used as working capital or spent on equipment purchases. It is growth capital that is essentially a loan that requires repayment in full, plus interest. IP-based securitization is debt financing that utilizes collateralization of IP to obtain a loan or for establishing credit.

13.3.1 What Does Collateralization Mean for My IP?

Intellectual property assets represent important, valuable corporate assets that can be used as a source of debt capital. Many traditional lenders are uncomfortable lending capital to unstable borrowers without some sort of collateral bearing value borrow which to against. Collateralization allows an IP rights holder to derive value from their intangible assets by securing financing that would otherwise be unavailable to them. A lender will distribute a loan to the IP rights holder in exchange for a securities right in the IP. In the event that the loan is defaulted on, the lender will assume the rights to the collateral IP.

The collateralization process is relatively simple to understand, but challenging to execute because the IP assets must be valued prior to lending. Intangible asset valuation for the purposes of collateralization is often a difficult undertaking for an IP appraiser because it requires not only an accurate determination of the fair market value of the IP and an accurate forecast of future revenues that could be generated by the IP being valued in the current context, but it also requires a prediction as to the theoretical future liquidation value of the IP in the event that the loan is defaulted on. Where normally IP assets are appraised at fair market value, which is the value that two willing parties would negotiate in an arms length transaction to purchase the rights, those IP assets that have been seized in the event of default will be liquidated by the lender under distressed circumstances, and will likely be worth less than fair market value.

The future liquidation value is challenging to determine since it is unclear what the worth of the IP will be at the time of default. If, for instance, the company's default on the loan is a result of the technology being a failure, the liquidation value of the collateral IP assets will be worth less than the liquidation value of the IP assets seized if the default is a result of poor company management. Little historic data is available on liquidation methods for intangible asset liquidation valuations, compelling IP appraisers to craft new, creative solutions for the valuation of IP.

There are many benefits to using IP assets as collateral for a loan. Collateralization provides a source of capital that does not dilute the company's equity structure. Risk management is another benefit to IP collateralization. In non-recourse financing the lender is only entitled to repayment from the profits resulting from the investment loan; the lender may not collect repayment from any other assets owned by the borrower. The risk associated with licensing royalty payments is transferred to the collateral holder because royalties generated from the collateral IP are paid to the lender for the duration of the loan, until repayment is completed. If a licensee fails to pay its due royalties on collateralized IP, the lender has a cause of action to pursue the licensee for failure to pay, instead of the borrower/IP rights owner.

13.3.2 Collateralization: Getting a Loan and Establishing a Line of Credit

Some banks will permit the use of IP as collateral when obtaining a loan or for establishing a line of credit. IP-backed lending is a fairly straight forward debt instrument. As a borrower, it is important to establish title to the collateral IP as a preliminary matter. No lender is comfortable making a loan in reliance of collateral when it is uncertain who exactly owns the rights to the IP, even if the borrower purports to own the rights. It is highly recommended that a filing be made with the USPTO regarding any collateral assignments of patents or trademarks.

After ownership is confirmed, a lender will be concerned with perfection of the security interests in the IP. Perfection is a notification process whereby the secured parties' rights become fully enforceable. Two sets of laws govern perfection of IP collateralization, federal law and Article 9 of the Uniform Commercial Code. Based on the type of IP used as collateral, perfection may require compliance with one or both. To perfect the security interests in patents and trademarks, compliance with the filing requirements of Article 9 is required. Perfection of copyright interests, on the other hand, may only require federal registration with the Copyrights Office. However, if the securities interest in the copyright also grants an interest in related receivables derived from the copyright, then compliance with Article 9 is also required.

Another creative IP-based financing solution is asset based lines of credit. These lines of credit are designed to provide a company with the capital necessary to bridge the gap between cash flows from receivables and its business expenses. The line of credit can be used for an assortment of business expenses from operating costs to working capital used to promote company growth, or as capital for turnarounds, buyouts, mergers or acquisitions. Where traditional lines of credit are based primarily on the creditworthiness of the borrower, asset based lines of credit are based on the value of the IP assets. Typically only IP assets with demonstrated market value are factored into the credit assessment of the assets. These could be IP assets that have previously been bought and sold, or those IP assets with licensing agreement in place which are known to generate revenue.

13.3.3 Case Study: Cambridge Display Technology Ltd. Gets $15 M IP-Backed Loan from Lloyds TSB

Cambridge Display Technology Ltd (CDT), a licensing company, held IP rights on fundamental patents that were central to the next generation of flat panel display technology. Using its IP position to its advantage, CDT sought to commercially introduce its revolutionary polymer-based organic light emitting diode (PLED) technology in 2004. CDT's licenses were scheduled to begin commercial production of the PLEDs in mid-2005, which would result in substantial royalty revenue for CDT.

Around the same time, CDT was looking to break away from its equity owners and was focused on becoming a self-sustaining business entity. As such, CDT needed funding to sponsor research and development projects until its royalty revenues on the new PLED production would be paid by licensees. Rather than dilute the equity in the company further, CDT sought an alternative source of working capital. In July 2004, CDT received a $15 million loan from Lloyds TSB, a corporate bank based in the United Kingdom. CDT's IP portfolio served as collateral for the loan through a credit enhancement firm, IP Innovations Financial Services, Inc. IP Innovations fully guaranteed the loan and secured the loan for CDT. IP Innovations guaranteed the loan based on:

- The strength of the CDT's patent portfolio
- CDT's extensive licensing history
- CDT's unprecedented up front licensing fees based on the fledgling PLED precommercial technology
- The size and growth trajectory of the market for the products supported by the PLED patents, and
- The degree of industry-wide investment towards commercializing the technology.

It is important to note that IP collateralizations are based on very conservative numbers when calculating the risk involved in the transaction. IP Innovations hedged its risk on the transaction by using conservative loan to value ratios and shortening the maturity length of the loan. IP Innovations wanted to ensure that timely remarketing of the IP would be possible in the event that CDT defaulted on the loan, thereby foreclosing on the note. IP Innovations shortened the loan repayment window to 3 years. Contemplation of these risks assessments were essential considerations to IP Innovations' decision when guaranteeing CDT's loan.

13.4 Blended Debt-Equity Financing

Another financing option involves a blend of debt and equity financing, also commonly referred to as mezzanine financing. Debt and equity financing are not mutually exclusive and are often combined to allay some of the disadvantages of each. Ideally, the blend of debt to equity financing should maximize return on the investment while minimizing risk. A mezzanine loan can either be a secured debt requiring collateral, or an unsecured debt, meaning the investor does not require collateral in exchange for the loan. For the secured debt scenario, some mezzanine investors will permit the use of IP rights as collateral. In situations where the debt is unsecured, the investor holds a convertible note which can be converted to equity in the company if the loan is not repaid. The result of a conversion is dilution of the ownership structure of the company, giving a stake in the company to former creditors. However, mezzanine financiers are not particularly interested in becoming shareholders in the company—they would prefer to be repaid on the loan. The convertible note is a fall-back measure in the event that the loan is not repaid.

13.5 IP-Backed Securitization

IP-backed securitizations are quickly becoming a popular means of debt financing. For instance, an IP-holder's future cash flows, or royalty payments, from future patent licensing agreements can be exchanged for an up-front, lump-sum payment by an investor. Essentially, the IP-holder receives cash in-hand and trades its right to royalty payments on future licensing deals, or a portion of the royalty payments, which may result from its IP for a set number of years. Securitization in this fashion puts the risk on the investor because the investor pays out a set amount to the IP-holder, which it may or may not recover as a result of future licensing deals.

Securitization is a mutually beneficial form of debt financing for both the investor and the IP-holder. There are two main models for securitization of future royalties: royalty interest transactions and the revenue interest model. In a royalty interest transaction, the transaction gives the IP-holder an up-front payment for its IP rights, while the investor attempts to buy the future revenue stream from the IP at a discounted rate, with the hope that the IP turns out to be worth more than its purchase price over the duration of royalty payments. If the investor is correct, the purchase could be well worth the risk.

A revenue interest model is similarly structured but is executed before the IP rights have generated any revenue whatsoever. Due to the lack of established earning potential of the IP rights and the greater level of risk associated with the IP, the investor can typically negotiation better terms for the transaction. The investor is effectively paying for all prospective royalty generation from the IP rights.

13.5.1 Case Study: Debt Financing with Bowie Bonds

In 1997, Bowie bonds, named after rock musician, actor, and record producer David Bowie, were asset-backed securities based on current and future royalty revenues that would be generated on songs recorded by Bowie prior to 1990. By that time, Bowie had amassed 25 albums, consisting of 287 songs, which he collateralized in a unique and pioneering securitization scheme. David Bowie was the first musician ever to securitize his IP rights in his music as collateral for a loan.

The David Bowie Class A Royalty-Backed Notes were given an A3 rating by Moody's Investors Services, based on a 7.9 % annual return, which at the time was a better rate of return than U.S. Treasury bonds. The Prudential Insurance Co. was eager to buy the bonds, and purchase the entire bond issue. Bowie had to forfeit his rights to all royalties for a 10-year period in exchange for an up-front lump-sum payment of $55 million. This transaction took the form of a loan. Over the 10-year period, the bond holder would get an annual principal payment plus interest. In the event that the contractually stipulated debt obligation was not satisfied by the royalties generated over the life of the bonds, Bowie would sacrifice his IP rights in his music to the bond holder. Prudential held on to the bonds and did not make them available for repurchase by individual investors on the secondary market.

Bowie bonds met the market with such initial success that other artists also opted to securitize their IP rights in their musical creations. The individual responsible for structuring the Bowie bond debt vehicle was a banker named David Pullman. Pullman assisted other artists securitizing their music with debt structures called Pullman bonds, which were similar to the Bowie bonds. Some notable artists include James Brown, the Isley Brothers, Ashford & Simpson, heavy-medal band Iron Maiden, Marvin Gaye and Rod Stewart.

Bowie bonds were downgraded to Baa3—near-junk status—in 2004, citing lower-than-expected revenues due to a slowdown in sales of recorded music. Illegal downloading was largely to blame for the slowdown, as file-sharing was rampant in the early 2000's. Apple helped to legitimize the downloadable music and media content industry with the popularization of Apple's iTunes and iPod products. Instead of having to buy an entire record just to own and listen to one or two songs, consumers could instead purchase the rights to individual songs. This change in the way people purchase music generated a revival of the music industry.

13.5.2 Case Study: Yale University's Monetization of IP Deal with Royal Pharma

Examples of creative monetization of IP rights can be found in many industries. Another example of IP monetization based on royalty financing can be seen in the pharmaceutical sector. The first notable, and highly lucrative, example of patent securitization involved a drug patent held by Yale University. In 1985, at the height of HIV and AIDS pandemic, Yale University secured a patent on a new drug for treating the HIV virus. Yale granted an exclusive license of the patented technology to Bristol-Myers Squibb, for further development of the drug called Zerit®. Zerit® became a commonly prescribed thymadine nucleoside reverse transcriptase inhibitor for use in HIV therapy. It was approved by the FDA in 1994, and Bristol-Myers Squibb made royalty payments to Yale for use of the patent rights.

In 2000, Yale University entered into an agreement with Royalty Pharma to securitize the royalty stream associated with the Zerit® drug. In exchange for an up-front $115 million payment, Yale forfeited royalty revenues to Royalty Pharma. The securitization vehicle, BioPharma Royalty Trust, was established to shield both Yale University and Royalty Pharma from bankruptcy in the event that the transaction ultimately failed. To finance the deal, the trust raised the $115 million: $57.15 million in senior notes, $22 million in mezzanine and $21.16 million in junior notes, with additional funding derived from $14.69 million in equity. This transaction was the first of its kind to obtain a single A rating from Standard & Poor's based on the credit ratings of Yale University and Bristol-Myers Squibb and the projected sales of the Zerit® drug.

Unfortunately, the deal ultimately failed when it entered into early amortization in 2002 as a result of lower-than-expected cash flow. Bristol-Myers had sold its entire IP portfolio related to the HIV drug to another entity, thus leaving Bristol-Myers with no reason to continue licensing the IP rights to the drug from Yale University. The Zerit®-based securitization model may have been a success if, for instance, the royalty revenues were not derived from a single drug, but rather a group of drugs. Learning from its mistakes with the Zerit® securitization model, Royalty Pharma again attempted to securitize patent royalty revenue in 2003 and met with great success. This transaction involved the revenue generated by a pool of thirteen drugs. This transaction was worth $225 million, obtained as a single tranche that was insured, and was rated a AAA rating by both Standard & Poor and Moody's Investors Services.

13.6 IP Insurance

13.6.1 Liability Insurance/Infringement Defense Insurance

Insurance related to IP is available in two main forms: (1) liability insurance; and (2) pursuance, or abatement, insurance. Liability insurance is typically defensive insurance geared towards fighting against lawsuits alleging patent, trademark or copyright infringement. Coverage typically reimburses the insured for legal expenses incurred during litigation and covers defense expenses, such as attorney fees, injunctions and appeals, arbitration and mediation expenses, the cost of asserting patent invalidity or initiating post-grant proceedings as a defense to the infringement action, and damages such as judgments, settlements, lost royalties, interests and costs. However, some policies exclude damage awards and cover only defense-related litigation expenses.

While IP insurance is still relatively new, its benefits are readily apparent. Defensive IP insurance, can provide an insured/defendant with sufficient funds to support infringement litigation from beginning to end, where without IP insurance, some defendants are forced to settle or submit to a licensing agreement because they cannot afford the litigation. Furthermore, defensive insurance is a source of litigation funding that does not come from the working capital of the insured entity. IP insurance can also attract investors, since any investment made will not be diverted into litigation expenses.

An IP holder seeking coverage is unlikely to find a willing insurer if there are any known allegations that the IP infringes the rights of another. Insurers commonly will not provide coverage to IP rights holders if there are any known, pre-existing threats of infringement allegations asserted against the IP holder, which can include things like cease-and-desist letters or emails, verbal warnings about the alleged infringement or other various forms of infringement notification. Furthermore, the insurer will not cover any past judgments, or any costs incurred during previous litigation. It is also common for IP insurance policies to include a 90 days exclusionary period once coverage begins during which any infringement action brought against the policy holder is excluded from coverage.

13.6.2 IP Abatement Insurance/Enforcement Insurance

IP abatement insurance, also called enforcement insurance, covers issued patents and patent applications, registered and filed trademarks, both registered and non-registered copyrights, as well as trade secrets. It is a plaintiff's policy, designed to assist the plaintiff/IP owner with enforcement of their IP rights against alleged infringers. The policy covers legal expenses associated with the enforcement of the insured's IP rights. An offensive IP insurance policy provides leverage in transactions and licensing negotiations because the policy reduces the financial pressure to settle prematurely. The policy improves the intrinsic value of the IP and demonstrates confidence in the enforceability of the rights, which in turn may discourage potential infringers from infringing on the IP rights. The insurance policy facilitates litigation of infringing activity by being a source of sufficient funding to fuel litigation efforts. Abatement insurance also typically covers the costs of defense litigation of counterclaims of invalidity, and the costs associated with relevant post-grant proceedings.

13.6.3 Typical Policy Terms

Policy terms may vary based on the type of insurance sought (defensive or offensive) and based on the insurer. Many terms are typically offered by those entities that provide IP abatement insurance. For instance, coverage can be obtained for domestic actions, foreign actions or both. Coverage duration is typically between one to three years. Policies are offered at a 1–3 % premium. For example, to obtain a $10 million policy at a 1 % premium would cost the insured $100,000 annually. Co-pay is often 10 % of the coverage amount, but can be increased to 20 % to reduce premium costs. Claim limits depend on the policy and the IP which is insured, but are often capped at $10 million, with higher limits being available to those entities that qualify.

Chapter 14
Licensing of Intellectual Property Rights

Licensing IP can be used to learn about and use others' technology. IP-holders may exclude others from using their protected IP. Licensing agreements are effectively grants made by the IP-holder to others that grant access to the protected technology and trade secret information, while creating a revenue stream for the IP-holder. For example, a high-technology consumer electronics company may need to license proprietary manufacturing equipment from others. Conversely, a company may consider licensing its product along with a trade name and marketing campaign to a third party to create a larger distribution network and generate revenue from a larger market. Finally, IP rights can be licensed or, in some cases, cross-licensed to resolve litigation.

14.1 What is an Intellectual Property License?

An IP license is a contract between two parties allowing the licensee to use at least a portion of the licensor's intellectual property in exchange for some consideration. Some examples of what the licensor receives are: a lump sum, multiple payments, royalty stream, goods, services, a cross-license to the licensee's IP, or combinations thereof. Where there are multiple payment types due during the term of the license, the payments may be tied to the sale of goods or services provided or sold (e.g., as a percentage of gross or net revenue for a product that the licensee is selling). Technology transfer agreements, which are often used by universities to capitalize on research and development, often involve an up-front minimum payment as well as a stream of royalty payments.

> *Example*: In 2005, the Semiconductor Manufacturing International Corporation (SMIC) was looking for a strategy to help it maintain its competitive edge in the non-volatile memory (NVM) market. Being a world-leading semiconductor foundry, SMIC needed the next step in NVM technology. SMIC found what it was looking for in Saifun Semiconductors' NROM®

G. B. Halt, Jr. et al., *Intellectual Property in Consumer Electronics, Software and Technology Startups*, DOI: 10.1007/978-1-4614-7912-3_14, © Springer Science+Business Media New York 2014

technology. Saifun's NROM® technology is used in the production of flash memory-based products and allows for storage of up to four bit-per-cell (over twice the storage capacity of basic memory cells). SMIC lacked the IP rights to this new technology and Saifun lacked the manufacturing capabilities to produce its patented technology. Thus, SMIC and Saifun entered into a licensing agreement where SMIC was permitted to manufacture NVM based on Saifun's patented NORM® technology and both parties benefitted from the arrangement.

14.2 Factors to Consider in an IP License

There are numerous factors that need to be considered in an IP license which are unique depending on the facts and circumstances presented and rarely, if ever, are two licensing deals the same. Such factors may include what rights to license, term, territory considerations, exclusivity, the amount and structure of the royalty, indemnification from damage caused by the other party, etc. Special consideration is also required if a license relates to the use of a licensor's trademark. These issues are discussed briefly below.

14.2.1 Identification of Rights to be Licensed

Patent rights include the right to exclude others from making, using, selling or offering for sale a patented article or process. By licensing a patent, the patent holder is giving permission to the licensee to no longer be excluded from practicing the patented invention. Patent rights are territorial, and are limited to the country in which the patent is granted. Trademark rights are the right to use a trademark in connection with goods or services, and are also territorial by country. Other rights can provide access to a proprietary or trade secret technology or know-how.

Example: Tyco Electronics produces many patented electronic technologies and components, which it has made available for sale either directly from its company website or through authorized distributors and resellers. While Tyco holds the IP rights to a particular electronic device that is for sale, it has granted the authorized distributors and resellers the right to sell its products.

When creating a licensing agreement, Tyco Electronics should identify the rights that are going into the license. This could include the rights of one or more of patents directed at the particular device or technology, as well as the know-how related to manufacture of the product, trademarks associated with the product, whether registered or unregistered, and possibly trade secrets, such as the product formula, if there is one.[1] As patent rights are for a limited term (the maximum life of a patent being 20 years from the earliest filing date) and any patent license would automatically terminate once the patent expires, it is beneficial to also license intellectual property rights that do not expire, such as a trademark and/or technology and know-how for manufacturing the product so that the license does not have a fixed term.

14.2.2 Restrictions

A license allows the licensor to restrict a licensee's activities to something less than an unlimited right to use the licensed intellectual property. For example, a patent licensor may choose to grant a licensee to make, use, sell or offer for sale a patented product.

Another restriction is whether the licensor is granting an exclusive, sole or non-exclusive license. An exclusive license is similar to an assignment of the IP rights to the licensee, and the licensor foregoes the ability to use its own IP rights in favor of the licensee. An exclusive license presents the most value to the licensee because it prevents all competition from using the licensed IP rights, and is often granted by research institutions that have no intention of commercially exploiting an invention. A sole license, in contrast to an exclusive license, allows the licensor to continue to use the IP rights, but limits the licensor from granting any further licenses to third parties. Thus, in a sole license agreement, both the licensor and the licensee may use the technology under license. A non-exclusive license allows the licensor to grant multiple licenses to third parties, which allows for multiple companies to use the licensed technology, which increases competition.

Various types of restrictions can also be included in a licensing agreement. For example, territorial restrictions are a common restriction in license agreements. A license should explicitly define the licensed territory or geographic area where the licensee may use the licensed technology. It is very common in licensing agreements for a company to only be permitted to service a limited geographic territory. In exchange for the territorial limitation, the licensor will not make a

---

[1] Divulging trade secret information, even under the terms and safeguards noted in a license, can involve the risk of loss of the trade secret, and often it is beneficial to structure the license so that a trade secret is not divulged.

similar grant to any third party allowing the third party to compete in this same restricted area. This is a very common practice in the operation of franchises.

Field of use restrictions are also frequently included in licenses, and may limit the use of the equipment or components contained within the product to a specific type. Field of use restrictions could limit the use of the licensed technology to certain applications such as therapeutic applications, veterinary applications, industrial materials applications, etc. This allows a licensor to grant multiple licenses in various fields of use, enabling the licensor to exploit its IP more effectively.

Product restrictions can also be included in a licensing agreement. This limitation restricts the licensee's use of the IP to a particular class of product. For example, a semiconductor company may license to a consumer electronics manufacturer the right to use its high-end, germanium semiconductors in its microprocessor products for use in personal computers, specifically laptops computers.

> *Example*: Incorporating a popular song into the background of a television commercial can help solidify the memory of the advertisement in a viewer's mind. If the viewer of the commercial has a positive association with the song, likes it, or thinks it is catchy, the viewer will be more inclined to remember the advertisement, associate positive feelings with it and may even look forward to seeing the commercial again. When a marketing company makes an advertisement and uses a popular song, the marketing company must procure a license to use the sound recording of the popular song. The label company that owns the song may put restrictions on the use of its song. For instance, the label may limit the use of the song to only commercials for a single product line or may require that the song only be used for tasteful advertisements.

14.2.3 Consideration

Consideration between parties to a contract regarding intellectual property rights can take many forms, but usually involves some form of payment. The most common are a lump sum payment, a stream of royalty payments (most commonly based on the number of products sold), as well as a combination of both. A lump sum payment can be used in a number of circumstances, such as settlement of past infringement, or when the technology is being transferred and the licensor is going to be working in the field. This type of royalty shifts the entire risk of success onto the licensee since the licensor receives payment no matter what happens. As a result, an up-front lump sum payment is often discounted from the amount that could be obtained by taking a stream of royalty payments over time based on the sales volume of the product. A stream of royalty payments over time based on the sales volume mitigates the risk that the licensee must take, because the licensee does not make any payments unless the product is selling. The licensor stands to receive a greater stream of royalty revenue if the product does well in commerce. A combination of some up-front lump sum payment to offset research and development costs incurred by the licensor along with a reduced stream of royalty payments is sometimes used to strike a balance for a licensor who wants some immediate payment, and also wants to share in the success of the product.

Many licenses are structured so that a minimum royalty payment must be made during a given time period in order to maintain the license in force. This prevents a licensor from obtaining, for example, an exclusive license and then failing to take action to manufacture, advertise or sell the product. Failure to meet minimum payments can act as a trigger for automatically changing a license from exclusive to non-exclusive, or for terminating the license.

Setting royalty rates is considered to be an art form more than a science. While some reference materials are available that provide "typical ranges" of royalty rates for a particular field, each situation is unique and must be reviewed independently based on all of the available information, such as the predicted market size, ramp-up time, whether production is capital intensive, etc. Additionally, the relative size and strength of the parties is often taken into consideration as well.

In addition to royalties, other non-monetary items can form the consideration. Cross-licenses may be obtained if both parties possess intellectual property rights that the other party desires. This may replace or offset some or all of the monetary consideration.

14.2.4 Maintenance of IP Rights

There are a number of terms that should be included in a license agreement that relate to the maintenance of the licensed intellectual property rights. For patent rights, this should include the requirement that the licensee include the appropriate patent marking on any goods sold that are covered by the patent. Lack of patent marking can limit claims for damages against infringers. Depending on whether the license granted is an exclusive license, the licensor may also require the licensee to pay patent maintenance fees. In some situations, a licensee may even take over prosecution of pending patent applications, which may benefit the licensor especially in the situation where the licensor is an individual or has limited resources to prosecute the patent application.

In trademark licenses, in order for the licensor to maintain its trademark rights, the agreement should require that any goods and services using the trademark identify that it is being used under license from the licensor. Additionally, to maintain the mark, the trademark license should have provisions for inspecting any products which use the trademark to insure that the quality of the goods is consistent with the licensor's standards. If a licensee does not maintain quality control over the usage of the trademark, not only can the licensor suffer damage due to poor quality goods being associated with the mark, but the trademark right can be lost.

14.2.5 Other Terms

License agreements will generally include a number of other terms, some of which are discussed below.

i. Representations and warranties are typically included from the licensor to the licensee, and should include a statement that the licensor is, in fact, the owner of the IP rights being licensed and has the right to grant the license. This offers some protection to the licensee from fraudulent transactions. The licensor should also warrant that they believe the IP rights to be valid, and should identify any known challenges to the IP rights. While a licensee should be conducting its own due diligence review of the IP rights being licensed, if the licensor fails to reveal this type of information, it can provide grounds for rescinding the license if the IP rights ultimately prove to be invalid.

ii. A license will often include terms regarding ownership and/or cross-licensing of further developments, in what is often termed a grant-back provision. From the licensor's perspective, this can be important if the licensor is also producing a product under the IP rights, and wants to have the benefit of any of the licensee's improvements, which are based on the licensor's underlying IP rights. From the licensee's perspective, having rights to improvements may eliminate the need for a further license relating to the same products or services being provided under the licensed IP rights.

iii. The burden for obtaining regulatory approval, for example, from the FDA, should be designated in the license. For new product types, this can be a time consuming process, and the burden for obtaining the required approvals should be specified. If the burden is placed on the licensee, this can be used as a negotiating point to obtain a reduced royalty, at least during the time it takes to obtain the approval.

iv. In the event that the parties ultimately disagree over the meaning or enforcement of any provision in the license agreement, some form of dispute resolution should be included in the agreement. This should not only include a choice of law, but a forum for any action or arbitration that will take place. In order to avoid the time and cost of a court proceeding, many agreements now call for binding arbitration of any dispute between the parties. A "loser pays" provision has also become standard in most license agreements to avoid meritless claims.

v. A termination clause is also standard in any license agreement, and should, in addition to setting any fixed or renewable term limits for the license, include a list of circumstances or actions that will result in automatic termination of the license. Automatic termination will generally occur for non-payment of any royalties due, failure to launch or market the product within a predetermined time limit, or bankruptcy of the licensee. Termination may also occur for breach of any other terms of the license agreement, generally after a notice and cure period.

A sample license agreement, which includes many of the above items, is attached in Appendix D.

14.3 Cross Licensing

Cross licensing occurs between two or more parties with symmetrical interests: A firm needs its competitor's patent just as badly as its competitor needs its patent. For example, the semiconductor industry utilizes cross-licensing frequently, with mutually beneficial results for the parties involved because the industry consists primarily of a limited number of players that produce similar products and hold similar IP portfolios. The same is true of the cell phone technology industry. Major players, such as Apple, Microsoft, Samsung, Motorola, Nokia and Research-in-Motion, are frequently in the news because of their newest IP portfolio cross-licensing deal.

Grant-back provisions in a cross-licensing agreement require the licensee to disclose and transfer back to the licensor any and all improvements that result from the licensee's use of the licensed technology during the licensing period. If there is no grant-back provision in the license agreement, the licensee could file improvement patents of its own, rendering the licensor's technology obsolete. It would be possible for the licensee to block the licensor from commercializing its own product if the licensee attempts to incorporate the licensor's improvements into the product.

Grant-backs can be exclusive, non-exclusive or an assignment. Under an exclusive grant-back, the licensor is granted the exclusive right to use or subli-cense any improvements created by the licensee. Conversely, the licensee retains merely a non-exclusive right to practice the improvements that it creates. In a non-exclusive grant-back provision, the licensee retains the title and rights to his or her improvements, but the licensor is allowed to practice the improvement as well. Under an assignment grant-back provision, the licensee must assign all rights and title of any improvements to the licensor. However, the licensee still retains a non-exclusive right to practice the improvement it was responsible for creating.

14.3.1 Case Study: Cross Licensing of Smartphone Technology: Microsoft and Samsung; Apple and Nokia; LG and Sony

Cross licenses are very common in the smartphone industry. When companies cross license patent portfolios, one party often supplements the license with extra payment, or boot, to cover any discrepancy in the value of the portfolio. Near the end of 2011, many of the large smartphone companies were signing cross licensing agreements.

Microsoft and Samsung

In 2011 Microsoft and Samsung signed a cross licensing agreement focused on patents relating to the Android operating software for smartphones. Prior to the agreement, Microsoft had been actively involved in making deals with other Android phone manufacturers, including HTC, Acer, ViewSonic, Velocity Micro and Winstron. Under those deals, the phone manufactures would pay a royalty to Microsoft per Android device sold.

While the details of the Microsoft-Samsung agreement have remained secret, the companies have both agreed publically that it was a mutually advantageous agreement. The cross license afforded both parties relief from all the litigation related to Android, as the suits were all dropped. In addition, Samsung agreed to coordinate with Microsoft on marketing and development efforts for the Windows Phone 7. Samsung also agreed to pay a royalty fee for each Android tablet and smartphone it produces and granted Microsoft access to its patent portfolio. Microsoft reciprocated by granting Samsung access to its patent portfolio.

Apple and Nokia

Apple and Nokia entered into a cross licensing agreement after engaging in a series of disputes in the U.S. and Europe between 2009 and 2011. The disputes focused on patents related to enabling wireless handsets to operate on GSM, 3G and Wi-Fi networks and also included patents related to wireless data management, data encryption, wireless security and speech coding. Nokia alleged that Apple was infringing its patents and sought royalty payments for each iPhone that was ever sold. Apple countersued claiming that Nokia infringed thirteen of its patents. Both companies also sought relief in parallel proceedings before the International Trade Commission. Entering into the cross license settlement eliminated all of the pending litigation between the parties. While the details of the settlement were not publically disclosed, it has been speculated that Apple paid Nokia more than $600 M as part of the cross licensing agreement, plus an estimated $11.50 per iPhone sold in royalties.

LG and Sony

LG and Sony agreed to drop their patent infringement lawsuits when they entered into a cross licensing agreement related to televisions, smartphones, PCs and cameras. Sony initiated the patent battle before the ITC, claiming that several of

LG's phones, including the Rumor Touch phone, were infringing on Sony's IP rights. LG filed a complaint with the ITC as well seeking an injunction against Sony's Bravia TVs and the PlayStation 3, claiming that these devices were infringing LG's IP rights on Blu-ray playback technology.

14.4 Licensing Standard-Essential Patents[2]

Standard-essential technologies are those technologies on which an industry standard is built upon. The driving force behind developing industry standards is to promote the interoperability and compatibility of products manufactured by multiple companies around a single technology. Some familiar examples of industry standards include VHS for video tapes, MP3 for digital music and Blu-Ray technology for DVDs. Standard setting is typically accomplished, and monitored by, a standard development organization (SDO). Participation in an SDO is usually voluntary, and cooperation with the SDO provides companies with access to essential technologies, allows them to introduce and champion new technologies, and enables them to compete in the industry.

Industry standards exist to enable devices manufactured by many parties to work together; like Ethernet for a laptop and HDMI for flat screen TV. Wireless standards provide the foundation for the explosive growth seen recently in cellular communications and smartphone technology. These standards are promulgated by industry organizations such as the Institute for Electrical and Electronics Engineers (IEEE), European Telecommunication Standards Institute (ETSI), and the Third Generation Partnership Project (3GPP).

In most instances, participation in the SDO also carries the obligation that any IP rights owned by the participant must be offered for license to those who would like to implement the standard. For example, ETSI, the SDO that produced the universal mobile telecommunications system (UMTS) and long-term evolution (LTE) wireless cellular standards, has an IP rights policy that requires participants to offer licensing terms for patents that are standard-essential that are "fair, reasonable and non-discriminatory" (FRAND). In its most basic form, a standard-essential patent is one that claims an invention that must be practiced in order to implement the standard. FRAND licensing policies help companies provide standard-essential technology to their customers without violating the IP rights of patent holders, and the policies also ensure innovators are compensated according to fair and reasonable terms for their contributions to the standard. FRAND policies also promote competition by lowering the barriers to entry into the market.

---

[2] Rob Leonard and Amber Stiles, excerpts from *The Current Smartphone Patent Landscape: The Importance of Standard-Essential and Implementation Patents*, New Jersey Lawyer Magazine Oct. 2012. This article was originally published in the October 2012 issue of New Jersey Lawyer Magazine, a publication of the New Jersey State Bar Association, and is reprinted here with permission.

FRAND licensing exists primarily to prevent an SDO participant from placing patented technology into a standard and then excluding others from implementing the standard by enforcing the exclusionary rights of the patent. By requiring fair and reasonable licensing of standard-essential technology, no one participant of the SDO can influence the deployment of a standard. The FRAND policies also promote competition by providing ground rules for licensing agreements and fostering a cooperative environment for licensing negotiations.

An interesting issue of contention surrounding FRAND licensing obligations involves a patent holder's ability to seek injunctive relief against those who infringe standard-essential patents subject to FRAND licensing obligations. On the one hand, if FRAND licensing was offered but no agreement was reached, one of the exclusionary rights afforded patent owners is the ability to exclude someone from making, using, or selling the patented technology. Injunctive relief is a stick by which a patent holder can force a license or other settlement. On the other hand, an SDO participant who inserts patented technology into a standard, feigns attempts at FRAND licensing, then seeks injunctive relief will appear anti-competitive and at odds with the intent of many SDO IPR policies.

The recent developments in SDO IPR policies and the uncertainty of injunctive relief have caused a shift as to where companies capture value from their patents. Standard-essential IPR, while still valuable as it must be practiced to implement a standard, is weighed down by FRAND and other IPR policy obligations of SDOs. In the current atmosphere, enforcement of patents on non-standard-essential technology—i.e., the other aspects of the device that are not part of the standard— is of increasing importance to a company's patent strategy.

Chapter 15
Patent Pooling and Synergistic Business Relationships

15.1 Patent Pools

Patent pools can be beneficial and can enhance business relationships in a synergistic way, i.e., the parts of the whole relationship interact in such a way as to produce a joint effect—a juxtaposition—that is greater than the sum of its parts acting individually. When entities work together, and pool their IP assets, innovation can flourish. It is important that the pooling of patents promote innovation and not deliberately stifle competition in the relevant technology market.

Patent pooling is a licensing model where several entities collectively aggregate their collective patent rights into a package of rights to be licensed under a single license. The patents included in the pool should be complementary in nature, meaning that each patented technology in the pool should be necessary to practicing the patents in the pool. No patent included in the pool should serve as a substitute for another patent in the pool as a licensee has no need to license two patents that could be substitutes of one another.

Patent pools are a desirable licensing structure because they achieve economies of scale by mitigating transaction costs associated with licensing individual patents from independent patent holders. Rather than enduring individual negotiation costs for each individual licensing agreement, a pool participant can purchase a single all-access license to the entire pool. In essence, a patent pool is a one-stop shopping experience that permits a licensee to procure all the related complementary IP necessary to practice innovation within the relevant technology. Operation of a patent pool is generally an accepted practice, so long as the pool does not stifle competition and does not lead to the collusion of multiple entities as to violate the antitrust provisions of the Sherman Act.

G. B. Halt, Jr. et al., *Intellectual Property in Consumer Electronics,*
Software and Technology Startups, DOI: 10.1007/978-1-4614-7912-3_15,
© Springer Science+Business Media New York 2014

15.1.1 Patent Pool Construction

The framework for a patent pool is relatively simply, with the major caveat being awareness of antitrust concerns. Patent pooling can be simplified to six steps.

1. First, it is highly advisable that anyone seeking to implement a patent pool should submit a proposal of the pool for preliminary review to the United States Department of Justice and the Federal Trade Commission (the "Agencies"). This review by the Agencies will help identify any potential antitrust complications that could arise as a result of pooling essential complementary patents.
2. Second, the proposal should contain a clear definition of the scope of the pool. Defining the scope will keep the pool technologically relevant and small in size. Voluntary participation in the pool is likely, because participants in the pool have amassed sunk costs in the research and development of the patented technology. These costs could potentially be recouped by collaborating with other patent holders to form a pool.
3. Third, pool participants should then analyze and evaluate which patents are complementary and essential for inclusion in the pool.
4. Fourth, the pool participants should seek an objective independent review of the patents to ensure the patents' relevance to the pool.
5. Fifth, the pool participants should develop a pro-competitive licensing structure that is based on the Agencies' Antitrust Guidelines for the Licensing of Intellectual Property[1] to mitigate antitrust concerns.
6. Sixth, pool participants should collectively decide on a royalties sharing strategy for the pool.

This patent pooling model is designed to navigate antitrust difficulties commonly encountered by patent pools. By integrating the Antitrust Guidelines into a patent pool's framework, those seeking to construct a pro-competitive patent pool proposal, intended to remain well within the antitrust "safety zone," can use this model as a guide to their pool engineering process. Pools are designed to entice those patent holders with relevant technology patents to participate in the pool by providing economic incentive to voluntarily join. Pool contributors are presented with an opportunity to recoup some of their sunken research and development costs through licensing fees and royalty payments.

---

[1] U.S. Dep't of Justice & Fed. Trade Comm'n, Antitrust Guidelines for the Licensing of Intell. Prop. § 2.3 (1995).

15.1.2 Antitrust Concerns

The Agencies have laid out their general antitrust enforcement policy in the *Antitrust Guidelines for the Licensing of Intellectual Property*. While the Guidelines are not law, they are designed to help individuals predict whether the Agencies will raise antitrust concerns regarding their business practices. For example, the classification of the relationship between the licensee and the licensor may raise antitrust concerns if certain licensing restrictions are proposed for the licensing agreement.

The *Antitrust Guidelines for the Licensing of Intellectual Property* indicate that a horizontal relationship between the licensor and licensee(s) exists when, in the absence of the licensing agreement, the two parties would have likely been potential competitors in the relevant market. On the other hand, a vertical relationship exists concerning activities that are in a complementary relationship; that is, one firm is a consumer of a technology supplied by the other and the two firms are not competitors in that particular market. When there is a vertical relationship between the licensor and the licensee(s), the Agencies will look for potentially harmful anti-competitive effects resulting from that vertical relationship on any horizontal relationships observed at either the level of the licensor or the licensee(s).

The *Guidelines* indicate that some licensing restrictions promote competition, while others do not. Field-of-use, geographical, and exclusionary licensing restrictions can be pro-competitive and allow a licensor to exploit its IP in an efficient and effective manner. The Agencies have also designated several types of licensing agreement restraints as *per se* unlawful because they are anticompetitive in character. These include, but are not limited to, blatant price-fixing, agreements to reduce output, or division of the customer market among horizontal competitors. Each restriction hints at collusion between the licensing parties in an attempt to eliminate competition in the marketplace.

15.2 Synergistic Business Relationships

Patent pooling can create synergism between the entities involved in the pool, creating benefits that are irreproducible if the entities acted independently. Devising an IP licensing arrangement with a strategic ally, or collective of strategic allies, can create an innovative synergy that can have a bigger impact for the companies than if they operated separately. These business relationships are crucial to the success of the entity and they are not all created equally. Some business relationships are better for your business than others; generating greater revenues, more referrals, significant royalties, etc. Synergistic business relationships exist in a myriad of configurations: business-business, industry-research entity, industry-university, business-university, etc. These mutually beneficial relationships are strategic in nature and should be nurtured and developed as strategic alliances add value to your business.

One of the main advantages to implementing a patent pool is that it can open up the IP landscape surrounding a particular technology, allowing for new innovation that was previously inhibited due to the possibility of engaging in infringing activity. If, for example, at one point in time a technology developer found itself stuck and unable to innovate due to the possibility of infringing the patent rights of another, participation in a patent pool—a license to practice the previously prohibitive patents—would promote further development of the technology by eliminating the fears of litigation. Contributors to the pool and licensees gain access to the collection of IP encompassed by the pool, which can enhance and advance the business or research agenda of the contributor or licensee by opening up new areas for commercial or scholastic expansion.

15.2.1 Case Study: Google and Motorola Mobility's Synergistic Business Relationship

In the late 2000's smartphone technology was flourishing. Major players such as Apple, Microsoft and Motorola were all competing for market share in this highly lucrative industry. Apple and Microsoft both were capable of developing the necessary hardware, as well as the software aspects of their smartphone products in-house. However, Motorola only had expertise in the handset manufacturing realm; Motorola needed a partner to help furnish its smartphones with a user friendly and very popular operating system. Motorola found a synergistic business relationship when it partnered with Google to develop the Android operating system for use in Motorola's new smartphones. Motorola's handset devices utilize the Android operating system, licensed by Motorola from Google, Inc. The Motorola-Google partnership seemed natural in light of the core expertise of each company; Google consistently demonstrates a high level of expertise in programming and software technology, while Motorola consistently demonstrates mastery in creating popular handheld devices (considering the success of the Droid X, Droid and its very popular predecessor the RAZR). By combining the efforts of each company, both Motorola and Google enjoyed the mutual benefits from their synergistic business relationship.

In 2010, Motorola spun-off its smartphone business as Motorola Mobility Holdings. The spin-off was largely the result of the success of the Motorola Droid and Droid X smartphones, which utilized Google's Android operating system. The spin-off allowed the new company to focus solely on developing the next generation of Android platform smartphones. Then, in 2011, Google acquired Motorola Mobility as a subsidiary company at $40 per share, making the total price of the acquisition $12.5 billion dollars. Google, in part, sought to enhance innovation efforts by gaining access to Motorola Mobility's patent portfolio.

The acquisition provided Google with various opportunities to extract synergistic value from the deal. Google has limited knowledge and expertise regarding hardware, yet Google is currently pursuing expansion into new technology areas where hardware is integral to the technology. For instance, Google has plans to provide fiberoptic Internet service to residential neighborhoods and is engaged in producing a line of Chrome Netbooks (light weight laptops). It is clear by the acquisition of Motorola Mobility that Google is looking to enter the smartphone handset sector as well. The acquisition of Motorola Mobility gives Google the tools to develop a smartphone device on its own.

A synergy is also noted in terms of the patents acquired in the deal. Google acquired Motorola Mobility's patent trove consisting of 17,000 patents. Motorola Mobility's patents are complementary to Google's existing IP rights. The acquisition was also partially motivated by Google's desire to create a defensive patent strategy. Earlier in 2011, Google made a $900 million dollar bid to acquire the patent portfolio of the then bankrupt Nortel Networks Corporation. Google did not win on its bid for the portfolio of 6,000 telecommunications patents because six competing entities (Apple, Microsoft, Research-in-Motion, Sony, Ericsson, and EMC) formed an alliance and purchased Nortel's portfolio for a whopping $4.5 billion dollars in cash. With thousands of telecommunication patents in the hands of Google's competitors, Google needed a defensive strategy designed to create a disincentive for others to sue Google as it made its way into the mobile business with its Android software. By securing Motorola Mobility's IP portfolio, Google created a war-chest of defensive patents, which will help shield Google from litigation.

Interestingly, despite all the reports that Google acquired Motorola Mobility's patents for defensive purposes, within 3 months of the acquisition being finalized, Google launched a patent suit against Apple based on seven patents from Motorola Mobility. Google sought an injunction against Apple from the U.S. International Trade Commission to prevent the import of iPhones, iPads, and Mac computers manufactured abroad. However, within 2 weeks of filing the complaint, Google withdrew its claims against Apple. While many speculated that the withdrawal was the result of settlement discussions, but Google's filing with the ITC dropping its claims indicated, rather, that Google had reconsidered its potential for success on its claims.

Chapter 16
Divestiture

In today's high-tech world, it is readily apparent that transactions involving IP assets require a combination of technical, marketing, business and legal expertise. These assets serve as both strategic and financial tools for growing a business. An industry-focused IP portfolio can be worth a substantial amount of money if properly managed and utilized. However, IP assets that are not contributing to the success of a company are sometimes ignored and neglected; these assets may not be achieving a good return on investment. Simply because IP assets are no longer relevant to a company's business plan does not necessarily mean the IP is worthless. Rather than ignore the underutilized IP, the company could monetize the asset through divestiture, removing the IP from the portfolio and converting the asset into cash.

16.1 Sale of IP

In situations where IP is no longer relevant to the business goals of a company, the company may decide to sell the IP rights. Since IP assets can be treated like any tangible asset, sales are often quick transactions, assuming that the valuation process for the particular asset is relatively straightforward. There is a whole industry dedicated to the sale of IP assets. There are some companies whose entire business model is centered on buying IP rights directly from the seller and then licensing those rights to others. A seller could also retain the services of a patent broker to assist in negotiating a sale.

A sale is usually a complete transfer of the rights to a new owner. However, sales agreements for patent rights have been known to include a license granted back to the seller/patent holder so that he or she may continue to use the technology covered by the patent in exchange for a small fee. This type of license is usually nonexclusive. Sometimes the buyer retains a right to buy the license back, thus excluding the original seller from practicing the patent any longer. This right is usually reserved for situations where the buyer has plans to resell the patent to

G. B. Halt, Jr. et al., *Intellectual Property in Consumer Electronics,*
Software and Technology Startups, DOI: 10.1007/978-1-4614-7912-3_16,
© Springer Science+Business Media New York 2014

another party. The buyer turned seller may want to execute its right to buy back the licensing agreement because it is often preferred that the sale of a patent be unencumbered by licensing agreements.

Example: In August of 2012, Nokia agreed to sell 500 of its patents and patent applications to Vringo, a mobile device software company. The portfolio consisted of international patents and applications, for which Vringo paid a hefty $22 million. Vringo made the purchase with the intent of monetizing the patents, yet Nokia put conditions on the purchase that limited Vringo's ability to capitalize on its investment, including the requirement that a 35 % royalty be paid to Nokia on any gross revenue generated by the IP sold, as well as placing limitations on (1) Vringo's use of those patents subject to licensing agreements implemented by Nokia prior to the sale and (2) Vringo's use of "essential" patents.

Examples of companies that act as direct buyers include:

- **Allied Security Trust** (AST) is a trust with members from several high technology companies, including IBM, Intel, Motorola and Phillips. The trust buys patents directly from sellers.
- **RPX Corporation** is a patent risk solutions provider. It acquires patent rights and for an annual fee, members can use the RPX patent aggregation as a defense mechanism to litigation.

Examples of IP brokers include:

- **Epicenter IP Group LLC** acts as a purchasing agent for its clients, contacting and negotiating the purchase of patents, while keeping their client's identity protected.
- **ICAP Ocean Tomo Private IP Brokerage** is known for its involvement in the sale of large or complicated IP portfolios during live auctions. Sellers can broadly market their IP in a competitive bidding environment.
- **Inflexion Point Strategy** (IPS) provides representation for technology companies in buying, selling, and investing in IP assets. IPS has developed a private sale process based on an auction model with multiple bidding rounds spanning 4–6 months.
- **Tynax** operates an online technology trading exchange whereby the client can either hire Tynax professionals to act as a broker on his or her behalf, or can post his or her own portfolio on the online trading exchange.

16.2 IP Auctions

Auctioning off IP is another divestiture option, with many benefits to both parties involved in the transaction. An auction provides the seller with the opportunity to set a minimum price for the IP; the reserve bid. Auction-style sales give the seller's IP marketing exposure and allows for the market to drive up the price. At the end of the auction, the seller can feel confident that he or she got the most for his or her IP as was possible. On the other side of the transaction, the buyer has informed, open access to the buying opportunity and is on equal footing with other prospective buyers. With an auction style sale, the seller gets market transparency and the benefit of price discovery. Auctions can be held online or live.

Example: Nortel Networks Corp. was a Canadian telecommunications company that went bankrupt in 2009. To settle its debts, the company sold its physical assets at auction for a total of $3 billion and also sought to sell its patent portfolio of 6,000 patents via an auction. In 2011 Google placed an initial bid of $900 million on Nortel's portfolio, but it wasn't enough. A few months later, an alliance of telecommunications companies, including Apple, RIM, Microsoft, EMC, Ericsson and Sony, won the auction with a group bid of $4.5 billion. What is noteworthy about this example is that 60 % of the money generated by Nortel's asset sales was derived from the sale of its IP assets; Nortel's intangible assets were worth more than its tangible ones.

Examples of some IP auctioneers include:

- **Ocean Tomo Transactions** was a well known IP auctioneering firm, hosting many live IP auctions around the world. Ocean Tomo Transactions joined forces with ICAP in 2009, forming the IP auctioneering venture ICAP Ocean Tomo.
- **IpAuctions** offers an online auction forum for the sale of intellectual property assets.

16.2.1 Case Study: Ocean Tomo Auctions off NASA Patents

Ocean Tomo is a pioneer when it comes to the concept of live auctioning intellectual property rights. Using a model based on live auctioning of multiple lots, Ocean Tomo created an open and public forum for the exchange of IP. The auction-style format simulates a sense of urgency and gives closure to IP transactions. It also gives the IP liquidity and introduces transparency into the transaction where none previously existed.

In March of 2012, The Innovative Partnership Program Office of National Aeronautics and Space Administration's (NASA) Goddard Space Flight Center in Greenbelt, Maryland announced that it would be auctioning off twelve patents of NASA-developed technology in a March 29, 2012 auction facilitated by ICAP Ocean Tomo. The auction consisted of patents related to robotics, artificial intelligence, software-development, industrial process control and wireless sensor networks, organized into three lots. NASA was not auctioning off a complete ownership in the patents; rather, it was auctioning the right to an exclusive license to use the technology. Since U.S. taxpayer money funded the initial research behind the NASA patents, auctioning off the right to an exclusive license allows for NASA to successfully commercialize the IP, which in turn improves the tax-payers' return on investment because the technology that was developed by NASA is introduced into the private sector where it can be innovated upon and put to good use. At the conclusion of the auction, NASA was able to fetch $75,000 for the lot of software development patents, yet the remaining two lots did not sell at the auction.

16.2.2 Case Study: Auctions as a Result of Bankruptcy: Kodak Patent Portfolio Auctions

Kodak is an American icon when it comes to implementing a successful business model. The company has been in existence for over a century and was the first company to put a camera into the hands of ordinary consumers back in 1900 with its introduction of the Kodak Brownie camera. Kodak's business model was simple; have a great marketing scheme and provide the general public with an affordable personal camera that requires a constant stream of film in order to operate. By selling both the camera and the consumable film media which was needed to take pictures, Kodak was able to capitalize on both fronts.

Kodak's marketing campaign focused on the deeply-rooted sentimental nature of people and played to American's love of nostalgia—a fondness of memories of a simpler time. With Kodak camera products, one could relive a moment by capturing it on film. Americans were encouraged to make it a "Kodak moment" and savor it for years to come. Not only were photographs meant for private viewing, but they could also be shared with friends and family. Photos were a social instrument; photo albums serve as memorials of the past.

Kodak dominated the U.S. camera industry for much of the twentieth century. By the 1970s, Kodak held nearly 90 % of the market share for the photographic film industry. Kodak was also earning large margins on the sales of film production consumables used during the printing process of film development. However, an invention that would ultimately destroy the Kodak company was developed in 1975 by one of Kodak's own employees—the digital camera.

Initially, Kodak sequestered the further development of the digital camera as it was a threat to its continued success in the photographic film industry. Yet, the shift to the digital era was out of Kodak's control. As computers became more ubiquitous, digitization was the destiny of modern photography. Kodak was quick to adapt, and by the early 2000s, Kodak had developed a line of digital cameras called EasyShare. The cameras were compatible with a printer dock: simply place the camera on the docking station, press a button and pictures would instantly print. But even producing a competitive product couldn't save Kodak from the demise of the photographic film industry. As digital camera technology further developed, film sales ultimately fell. By 2006, Kodak was starting to fall apart.

By 2006, Kodak started engaging in a series of business operations divestitures and sell offs. Kodak divested its digital camera manufacturing operations to Flextronics in 2006. In 2007, Kodak negotiated a deal to sell off the Kodak Health Group to Onex Corporation for $2.35 billion, with the possibility of up to $200 million in additional future payments and also sold off its light management films business to Rohm and Haas. In 2009, LG Electronics acquired Kodak's organic light-emitting diode (OLED) business. 2012 saw Kodak sell its Image Sensor Solutions division to Truesense Imaging Inc.; its Kodak Gallery, a web-based photo management tool, was sold to Shutterfly for $23.8 million; Kodak quit its inkjet printer business; and filed for Chap. 11 bankruptcy, despite that all of these business division sell offs resulted in approximately $4 billion capital generation for Kodak.

Being located in Rochester, New York, Kodak filed for Chap. 11 in the United States Bankruptcy Court for the Southern District of New York. In its bankruptcy filings, Kodak claimed to have $6.75 billion in liabilities and only $5.1 billion in assets. With its physical assets dwindling and its debts outstanding, the bankruptcy court, in an order dated July 5, 2012, authorized Kodak to sell its IP at auction under the court's supervision. Kodak has approximately 22,000 patents and trademarks, with nearly 9,000 of those in the U.S. The auction's focus is on Kodak's digital imaging patent portfolio which consists of roughly 1,100 patents. The court originally set an auction deadline of August 13, 2012, but Kodak and its creditors agreed to extend the deadline when initial bids on the digital imaging portfolio did not generate the demand expected by Kodak analysts. Within days of the original auction deadline, Kodak announced that it might not actually sell off any of its IP through the court ordered auction process. Instead, Kodak began focusing on restructuring the company, in hopes that a restructure might save it from financial ruin. Kodak continued to sell off non-core areas of its business and negotiated a settlement with Kodak retirees to terminate its health-care and survivor-benefits program in an effort to resolve $1.2 billion retiree-benefits liability on its books.

However, by the close of 2012, Kodak ultimately sold its digital imaging IP portfolio at auction. The buyer was Intellectual Ventures Fund 83 LLC. In a concerted effort between Intellectual Ventures and RPX Corporation, twelve new licensees were organized to provide a portion of the acquisition funds, with the remainder paid by Intellectual Ventures. Intellectual Ventures received the portfolio of patents subject to the new, and existing licensing agreements. The twelve licensees include Google, Samsung, Facebook, Huawei Technologies, Microsoft,

Apple, Adobe Systems Inc., Shutterfly, Amazon Fulfillment Services, HTC, Research-in-Motion, and Fujifilm. Not only did the sale generate an influx of capital into Kodak, but also as a term of the sale, any current patent-related litigation between Kodak and any member of the purchasing/licensing consortium was required to be settled.

16.3 Cross Industry Applications: Channel Programs

There are some situations where patentable inventions may have applications in another technology area, which should be considered and explored as it could present viable commercialization opportunities. For instance, when contemplating the sale of a non-core IP asset, it is worth investigating whether the patent could fetch a higher sale price in a different industry. The patentable technology can have useful applications in adjacent markets or could exhibit a wide scope of applicability across a variety of industries. Consider the following example. A microelectronics company holds a patent on an invention relating to the fabrication process for producing nanowires (wires that are nanometers in diameter) for use in microscopic electronic circuitry. A method patent of this nature could also be used for fabricating nanofibers from natural or synthetic materials, assuming that the method for fabricating the nanowires is not limited to the use of conductive materials in the fabrication process. Nanofibers are used in many industrial applications including textile manufacturing, materials science applications, filtration media and biomaterials. A patent with such a wide impact in terms of industrial applications is highly valuable.

Similar to how channel programs work in business, establishing a channel program to sell off IP is another divestiture strategy that can prove highly lucrative. For example, a company seeking to shed non-core IP assets could establish a relationship with a non-practicing entity. Non-practicing entities typically build a business around a particular technology area. By partnering together in a channel program partnership, a symbiotic business relationship forms between the company and the non-practicing entity that is mutually beneficial. The non-practicing entity acquires fully developed IP rights, which the non-practicing entity could then either seek to litigate those rights or could license the technology, while the company generates income from its non-core assets.

16.4 Spinning-off IP

Innovative companies invest heavily in research and development of new technology. Sometimes the result is new developments in an area that is not the core focus of the company. While the ideas may be viable, commercialization of those ideas may not be feasible or in alignment with the business goals of the company. If the innovative technology lies outside of the company's core focus, a spin-off company could be formed around the new technology. Alternatively, a spin-off could also be formed in the event that a company is looking to exit a particular line of business. A spin-off based on IP typically has the following characteristics:

- **A significant investment.** There has been a significant corporate investment and several years of research and development expenditure on the technology covered by the IP.
- **A complete design.** There is a complete product or prototype in an advanced phase of development ready to be commercialized.
- **An eager market.** There has been an expression of interest in the product from consumers in the industry.
- **Knowledgeable employees.** Key employees, such as engineers or scientists familiar with the technology, were available as part of the spin-off.
- **Short time to market.** The product is close to launch and is within a year of entering the market and generating revenue.

Example: Agilent Technologies develops test and measurement tools and equipment for use in engineering, scientific, business and government applications. Agilent was created as the result of a spin-off of Hewlett-Packard in 1999. Hewlett-Packard created Agilent because the measuring equipment products were commercially viable, but were not related to computing technology, which is Hewlett-Packard's core focus. HP had made a substantial investment in its measuring equipment division, complete with product lines and a growing market to sell to. HP employees became Agilent employees since they possessed the skills and knowledge required to continue producing new and improved measuring devices. Manufacturing optical networking devices, biological, chemical and electronic test equipment, scientific instruments, semiconductors, Agilent has remained a highly successful spin-off company.

Examples of companies that assist with IP-based company spin-offs include:

- **Blueprint Ventures** is a technology investment firm specializing in IP-based company spin-offs. They assist small entities and entrepreneurs in building a successful new enterprise.
- **CRG** focuses on creating spin-offs based on maturing technologies by tapping into emerging markets. CRG follows a business creation model that takes spin-off companies from the research and development phase all the way through product expansion.
- **New Venture Partners** works with corporations to cultivate attractive investment opportunities based on technology that could be spun off as a successful and sustainable start-up company.

Chapter 17
Enforcement and Infringement of Intellectual Property Rights

17.1 Policing Your IP Rights

After a company invests the time and resources into developing and protecting its intellectual property rights, it is important to police the market to ensure that no competitors are improperly benefiting from such intellectual property investments. Policing of intellectual property rights can be broken down into a four step process.

(1) The rights must be identified by class: trademark, trade secret, copyright, or patent. It is also important to remember that any single right may be protected by more than one type of intellectual property protection. For instance, a patentable or trade secret right may have an associated trademark, or there may be copyrightable materials that accompany distribution of the patented item. Copyrighted instructions may be an important element in the commercialization of a trade secret. Consider the sale of a trade secret formula where the temperature, rate of mixing, and amount of the intermediary is critical to its performance in the end product. While the owner of such information would not want the disclosure to reveal the trade secret, the copyrighted instructions could be an important sales tool because it may be the only way to make the formula commercially useful to the customer. This identification process should be conducted carefully to consider all possibilities for protection.

(2) The protection of intellectual property rights identified through the prior analysis will require consideration of the competing marketplace of interest. If the identification step yields a product having commercial interest that is potentially patentable and marketing has identified or coined a trademark for the product, there are competing interests associated with the different forms of protection. On one hand, the patent interest requires diligence to be sure that the invention is not disclosed or sold more than one year prior to applying for the patent. On the other hand, trademark rights are based on use and there is a desire to initiate commercial use of the mark in association with a product as

G. B. Halt, Jr. et al., *Intellectual Property in Consumer Electronics, Software and Technology Startups*, DOI: 10.1007/978-1-4614-7912-3_17, © Springer Science+Business Media New York 2014

soon as possible. As noted in the trademark Chapter of this book, it is possible to file an "Intent to Use" trademark ("ITU") application before there is any disclosure of the invention or use of the trademark. There is a similar procedure, known as a provisional patent application, which can be filed to preserve a patent application filing date prior to any disclosure. Use of these two vehicles permits the preservation of patent and trademark rights while efforts are undertaken to gauge the market's commercial interest. This relatively simple solution to the potential problem is made possible through diligence in the identification of the concerned rights and remaining mindful of the critical first date of the underlying right.

(3) Vigilance in the marketplace is a critical component of intellectual property management. Market vigilance is the key to gathering information about competitive practices and products. This information is useful both for detecting infringement of rights and for avoiding infringement of third party rights. With regard to the issue of detecting infringement, the intellectual property owner is charged with a certain level of vigilance and long delays in detecting an infringement may result in an infringer having an equitable defense against the assertion of the infringed intellectual property right. Thus, you cannot sleep on your rights to the detriment of another. Conversely, one cannot count on ignorance of another's rights as a complete defense to a charge of infringement. As a general rule, one must act in a reasonably prudent manner with respect to its own rights and the rights of another. Frequently, a sales force or distribution network is an excellent source of competitive information. These individuals continuously interface with customers and interact in the marketplace, and are generally aware of competitors' products and services. While the law does not require the engagement of an investigator to search out all possible infringements, it does require an increased level of vigilance once there is a reason to believe there is an infringement.

(4) Enforcement is the end result of properly conducting the above steps and identifying an infringer. Enforcement is a process that should not be taken lightly, as it can have consequences for rights being asserted and consequences for the business entity itself. Once an infringer is discovered, there are steps that need to be taken for the purposes of evaluating the infringement and the impact on the business of the intellectual property owner. In other words, knowledge of an infringement carries with it the requirement of action or at least an informed decision not to take action. The steps needed for the purposes of evaluating the infringement and the impact on the business of the intellectual property owner should be followed in either case.

17.2 Evaluating a Controversy Prior to Commencing Litigation

A number of inquiries should be made prior to commencement of filing suit to enforce intellectual property rights. First, an initial assessment should be made as to whether the intellectual property right can be enforced. This involves identifying the right and checking its validity and enforceability. This evaluation needs to be made even when all steps have been taken to obtain a valid enforceable right, in order to ensure that the right has maintained its enforceability. For example, a patent may have lapsed for failure to pay maintenance fees and will no longer be enforceable under this scenario. Likewise, a trademark that has become abandoned (i.e., the owner is no longer using the mark and has no intention of resuming use) is no longer enforceable. In cases where an intellectual property right has been lost or become unenforceable, it can subsequently be regained or made enforceable again. For example, a patent that has lapsed for failure to pay maintenance fees may be reinstated up to 2 years after the lapse if it can be shown that the delay in payment was unintentional. With any intellectual property right, a check should be made to ensure that nothing has occurred to cause the right to cease being enforceable, and if it is found that such an event has occurred, a further check should be made to find if any steps can be taken to reinstate the right.

All elements of a legal cause of action must be present for successful enforcement. This evaluation is different for each type of intellectual property right.

- For a patent, it involves asking whether an allegedly infringing device meets each and every limitation of one or more of the patent's claims.
- For a trademark, it involves asking whether there is a likelihood of confusion between the asserted trademark and the allegedly infringing mark.
- For a copyright, it involves asking whether there was an actual copying of the work or a portion of the work and whether there is any evidence that the infringer has access to the copyrighted work.

With any cause of action, the first step is a check of whether all the elements of the potential claim can be satisfied.

It is always necessary to make a risk–benefit analysis of the economic impact of any potential litigation. Litigation can be costly, and the potential cost of a lawsuit is often difficult to predict. The American Intellectual Property Law Association recently published a report which found that legal fees in patent litigation could be $650,000 where one million dollars was at stake, and $2.5 million where more than one million dollars was at stake.[1] The estimated litigation cost should be weighed against the damages sought, and the probability of actually obtaining such an amount. It should be kept in mind that in many cases, remedies other than

---

[1] American Intellectual Property Law Association [1].

monetary damages may be awarded. A valuation of all potential remedies should be made. For example, if an injunction to stop an infringement is the main relief sought, an analysis of the value imparted to the intellectual property owner by the cessation of the infringing behavior needs to be made before any action is taken. That analysis should include the non-monetary costs of the time and energy that litigation takes away from the normal business operations.

17.3 Remedies and How to Achieve Them

17.3.1 Injunctions

Injunctions are a remedy in most intellectual property lawsuits, especially patent and trademark lawsuits. Injunctions can be sought in addition to monetary damages. As discussed in more detail below, an injunction can be permanent or preliminary.

A party seeking permanent injunctive relief must demonstrate by competent evidence that:

(1) It has suffered an irreparable injury;
(2) Other remedies available at law, such as monetary damages, are inadequate to compensate for the injury;
(3) The balance of hardships between the plaintiff and defendant warrant the granting of an equitable remedy; and,
(4) The injunction will serve the public interest.

Although permanent injunctions are more common in intellectual property lawsuits because monetary damage amounts are often difficult to ascertain, they are not considered to be automatic upon a finding of infringement. Injunctive relief in an intellectual property matter must meet the same four elements noted above.

17.3.2 Payment of Royalties

One possible remedy for patent infringement is an order that the infringer must pay the intellectual property owner a reasonable royalty. The court may determine what constitutes a reasonable royalty and order payment for past infringement, and, in certain cases involving public interest, order that royalties be paid for the future use of another's intellectual property right. An order for payment of reasonable royalties may be desirable where the actual damages are difficult to evaluate. Courts will often engage in a weighing of the harms when determining an appropriate remedy in intellectual property cases, and avoid awarding a remedy that will impose an undue burden on the infringer, particularly where the infringement was innocent. In some cases, courts will give an innocent infringer an option between an injunction and an order to pay reasonable royalties until such time as the infringer phases out the use of the owner's intellectual property without suffering excessive damage.

> *Example*: In 2009, Carnegie Mellon University sued Marvell Technology Group Ltd. claiming that Marvell was selling chips that infringed on two patents held by the University. Marvell counter claimed that the patents were invalid. The case was heard by a federal jury in Pittsburgh, Pennsylvania in late 2012. The jury returned a verdict in favor of Carnegie Mellon, finding that the patents were valid and that Marvell had literally and willfully infringed the University's patents. The jury also made a damages award of $1.17 billion to the University. This award was based on a reasonable royalty of $0.50 for every infringing chip Marvell sold worldwide since March 6, 2003. The $1.17 billion award is one of the largest verdicts ever granted. Marvell filed several post trial motions including a challenge to how the damages were calculated and a motion for a mistrial. If the post trial motions are successful, the jury verdict may be overturned. If Marvell's post trial motions are unsuccessful, it is likely that Marvell will appeal.

17.3.3 Monetary Damages

Monetary damages can be difficult and expensive to prove in intellectual property litigation because it is often difficult to ascertain the actual loss attributable to and incurred as a result of the infringing behavior. An investigation should be made to determine what damages were incurred, and if they are recoverable. For example, in a suit for trademark infringement, the trademark owner may seek damages for loss of goodwill resulting from the infringement, but measuring those damages can be difficult and involve surveys and experts that may push the costs beyond what is recoverable.

In certain cases, such as infringement of a registered copyright, the owner of the copyright may be entitled to statutory damages without the requirement of proving actual damages if the statutory requirements are met. The copyright statute provides, upon election by the copyright owner, that the court may award anywhere from $750 to $30,000 for each work infringed upon and may further increase the award per infringement up to $150,000 for willful infringement.

17.4 Settling Controversy without Litigation

17.4.1 Arbitration

Arbitration is a non-judicial form of resolution where the parties select an arbitrator or panel of arbitrators who hear evidence and decide the matter. The proceeding is similar to, but not as formal, as a court trial. The parties can agree that an arbitration award can be binding and enforceable in court and appeals may be permitted.

Arbitration offers various advantages over litigation, and several particular advantages unique to resolution of intellectual property disputes. Arbitration presents the opportunity to have a dispute settled in a more expedient manner than conventional courtroom litigation. Litigation of cases involving intellectual property issues can be long, tedious and complex. Arbitration may shorten the dispute process, and, because of the shortened time frame and limited need for discovery, reduce the costs compared to those typically associated with litigation.

One particular advantage of arbitration with respect to settlement of patent disputes lies in the fact that the arbitrator is chosen by both parties, permitting selection of an arbitrator having a technical background that will facilitate understanding of the patented subject matter. In 1983, the United States Patent Act was amended to provide for the voluntary settlement of patent disputes by binding arbitration.

Example: In early 2012 STMicroelectronics and NXP Semiconductor Netherlands BV entered into arbitration proceedings organized according to rules set forth by the International Chamber of Commerce. A claim was asserted against STMicroelectronics for underloading charges associated with the price of wafers that were supplied by NXP to STMicroelectronics's wireless joint venture during the period spanning between October 2008 and December 2009. The arbitration proceeding resulted in a ruling by which STMicroelectronics was ordered to pay $59 million to NXP Semiconductor. Additionally, the tribunal reserved certain issues raised by STMicroelectronics to be addressed by a second arbitration proceeding scheduled for June of 2012, with final resolution by June of 2013.

17.4.2 Mediation

Mediation can be an effective means of settling a dispute more quickly and at a lower cost than litigation, but it is very different from arbitration. The mediator functions differently than a judge, jury, or arbitrator by working with the parties to assist in a negotiated agreement that satisfies the interests of all without any determination of who is right or wrong.

Settlement of disputes by mediation offers several practical advantages. Litigation costs, such as discovery and motion practice, can be greatly reduced or eliminated. Another significant advantage is the abbreviated time lapse between commencement and resolution of the dispute, particularly due to the elimination of appeals which can draw out the litigation process. This is of particular importance in the context of patent litigation, where the right could expire or the technology involved could potentially become obsolete before the dispute is resolved. In addition, parties using mediation for dispute resolution avoid the risk of a complete loss on all counts, and, hopefully, negotiate a resolution that favors the continuation of the business interests of the party.

Mediation can be particularly valuable in resolving disputes over intellectual property rights because it offers the parties an opportunity to come to resolutions that may be unlikely to be obtained in court. Agreements may be reached that allow all parties to exercise an intellectual property right with minimal intrusion on the rights of the other parties. For example, in a dispute over trademark rights, the parties may come to an agreement where each agrees to keep limited use of the trademark, such as by confinement to a particular geographic boundary.

17.4.3 Case Study: NTP Patent Settlement with AT&T, Verizon Wireless, Sprint Nextel, T-Mobile, Apple, HTC, Motorola Mobility, Palm, LG, Samsung, Google, Microsoft and Yahoo

NTP Inc. is a patent holding firm, meaning that it is an entity that collects patents and enforces those patents against one or more alleged infringers. NTP does not practice the patents which it holds—it does not manufacture any products or provide any services. This type of entity is called a non-practicing entity, or NPE. NPEs will be discussed in more depth in Chap. 20.

In 2001 NTP sued Research-in-Motion (RIM), developer of BlackBerry devices, for infringing several of NTP's patents related to wireless emailing technology. By 2003 a jury found that the NTP patents were valid, that RIM had infringed them, and that the infringement had been "willful." The Judge ordered an injunction against RIM to cease its US sales of BlackBerry devices and services for the duration of NTP's patent protection, which was scheduled to terminate in 2012. RIM appealed and the court stayed RIM's injunction pending the outcome of the appeal. Ultimately in 2006, NTP and RIM negotiated a $612.5 million out-of-court settlement. The settlement included a licensing agreement that allowed RIM to continue its BlackBerry wireless operations.

After the successful settlement was reached with RIM, NTP began to assert its wireless email patents in various suits launched between 2007 and 2010 against thirteen technology companies including wireless carriers AT&T, Sprint Nextel, T-Mobile, Verizon Wireless; smartphone manufacturers Apple, Motorola Mobility, Samsung, LG Electronics, Palm, HTC; and email service providers Microsoft, Yahoo and Google. The suits involved eight patents held by NTP related to the automatic delivery of email over wireless systems.

During the litigation in the courts, parallel proceedings were taking place at the USPTO. The tech companies sought reexamination of the patents by the USPTO with the hopes that the USPTO would find the patents to be invalid. The USPTO did invalidate several of the patents, but NTP sought review of the USPTO's finding by appealing to the Federal Circuit Court of Appeals. The Federal Circuit overruled the USPTO's invalidation of NTP's patents. Within a year of the Federal Circuit's ruling, NTP and the thirteen technology companies entered a settlement agreement. Whether the Court of Appeals ruling triggered the thirteen tech companies to engage in settlement negotiations with NTP is up for debate.

While the precise terms of the settlement were not made public, it was made clear through a press release by NTP that the entirely of NTP's portfolio was licensed to the thirteen technology companies and that the agreed upon price left those on either side of the transaction satisfied. NTP received the compensation it deserved and the tech companies obtained a license on the whole of NTP's IP portfolio at a reduced rate. This settlement marks the first time that the majority of a single industry gained access to a patent holder's IP portfolio in a one-time settlement agreement.

17.5 Litigation

In cases where there is no other solution, litigation may be the necessary means to resolve the dispute.

17.5.1 Selecting a Jurisdiction

Intellectual property lawsuits may be brought in either the state or federal courts. The court used can be dependent on the issue and the authority for the asserted property right. Since trade secret rights most typically arise under state law, those suits are most common in state courts. Patent and copyright disputes are both the subject matter of federal statutes which invoke a federal court jurisdiction. Trademark rights may arise under state or federal law and the authority for such rights will determine whether the suit should be brought in state or federal court. Although most cases involving patent and copyright issues will be brought in federal court, there are exceptions where an intellectual property issue arises collaterally with a state law matter such as interpretation of a license or contract.

In addition to determining whether state or federal court is the appropriate place to bring suit, the defendant must reside in or do business within the state where the suit is brought.

17.5.2 Causes of Action

Infringement

Infringement actions are the most common type of litigation involving patents, trademarks and copyrights. An infringement suit may be brought for either direct or contributory infringement. In a lawsuit brought for direct infringement the party allegedly engaging in the infringing activity is named as the defendant.

For patents, infringement results from making, using, selling, offering to sell, or importing into the United States an invention within the scope of the patent claims without the patent owner's permission. For trademarks, infringement results from the unauthorized use of the same mark or a mark that has a likelihood of creating confusion between that mark and the plaintiff's mark. For copyrights, infringement results when the infringer had notice of, or access to, the copyright work and engaged in actual copying of all or part of the copyrighted work.

In some circumstances, the infringement may result from activity that is not directly infringing, but it encourages or enables infringement by a third party. In these cases, the intellectual property owner may bring a lawsuit against a party for inducing or contributory infringement (i.e., indirect infringement). Such causes of action for indirect infringement can be based upon various activities. For instance, the defendant may be making or selling product that is intended for an infringing use or is sold with instructions that provide directions for another to infringe. This situation is common in both the patent and copyright contexts. In either situation, it may be undesirable to file suit against the direct infringers as there may be a large number of defendants or the defendants may be customers, and the direct infringers are typically less likely to have recoverable funds than the indirect infringer.

Indirect infringement frequently arises where a patent contains only method claims. Someone making or selling a product that does not itself infringe the patent claims may encourage its use in a way that does infringe the claimed method. The patent owner may not be able to sue the maker or seller for direct infringement of the method, but could sue the maker or seller for indirect infringement if those activities induced others to infringe, or contributed to others infringing a patent.

Declaratory Judgments

A party having a real interest to capitalize on subject matter that is alleged to be covered by another party's intellectual property right may file suit seeking a declaration that there is no infringement of the intellectual property right in question, the right in question is invalid, or that the right in question is not enforceable.

The decision to seek a declaratory judgment requires serious thought. The owner of the rights in question may not pursue the potential infringement based on its own policy reasons, or may no longer have a sufficient commercial interest in them to expend the costs to pursue litigation. Thus, any decision to file a declaratory judgment action must under go the same analysis applied to any other litigation.

Trademark Dilution

Trademark dilution originated as a state law cause of action pertaining to common law trademark rights. In 1995, it was codified in the Federal Trademark Dilution Act and made applicable to federally registered trademarks. Dilution actions can only be brought by owners of famous marks. To make out a cause of action, the plaintiff must show that the defendant adopted the use of a mark that tends to dilute the distinctive quality of plaintiff's famous mark. If such a showing is made, the plaintiff can obtain an injunction against the defendant's use of the mark, even absent a showing of likelihood of confusion.

Misappropriation of Trade Secrets

To bring an action for misappropriation of trade secrets, a plaintiff must show the existence of a trade secret, and that the defendant misappropriated the trade secret. Under the Uniform Trade Secrets Act (UTSA), which many states follow, a trade secret is any information having independent economic value due to its nature and the fact that it is not commonly known, for which reasonable efforts are made to maintain secrecy. Misappropriation occurs where the defendant either acquired the information by improper means, or received the information with knowledge that it was derived by improper means or disclosed in breach of a duty to keep secret. Trade secret law does not take the "strict liability," approach of patent and trademark law, and requires some wrongful behavior on the defendant's part in order to impose liability. It should also be kept in mind that a party who "reverse engineers" the trade secret cannot be held liable for misappropriation of trade secrets, as this is not considered an "improper means."

17.5.3 Preliminary Injunction

Seeking a preliminary injunction at the outset of the litigation may be desirable to prevent further infringement during the proceedings. Such an injunction is temporary, and is typically only awarded upon a showing that the plaintiff has a strong likelihood of success on the merits, along with the potential to suffer irreparable injury if the allegedly infringing activity is permitted to continue during the proceedings. This can be a difficult burden for a plaintiff, particularly in intellectual property lawsuits where the court is often charged with detailed factual evaluations or subjective multifactor tests. Nonetheless, a case can be presented that warrants a preliminary injunction in order to mitigate damages.

17.5.4 Discovery

Discovery is a highly important phase of intellectual property litigation and its focus will vary depending on what type of intellectual property right is at issue. The Federal Rules of Civil Procedure govern intellectual property proceedings in federal court, while state rules will govern those cases that are brought in state courts.

Depositions, interrogatories, requests for admissions, and document requests are all available as discovery tools in intellectual property litigation. The Federal Rules of Civil Procedure provide for discovery of "any non-privileged matter that is relevant to any party's claim or defense".[2]

17.5.5 Summary Judgment

Motions for summary judgment can be an efficient means for bringing litigation to an early conclusion. Federal Rule of Civil Procedure 56(c) states that a party is entitled to summary judgment where "the pleadings, the discovery and disclosure materials on file, and any affidavits show that there is no genuine issue as to any material fact and that the movant is entitled to judgment as a matter of law."

Having a motion for summary judgment granted can be challenging in intellectual property cases where a large number of complex issues are often present and the evidence may be subject to different interpretations. One example is the case in which a defendant moves for summary judgment holding that the plaintiff's intellectual property right is invalid or unenforceable. While there are some instances where the controlling law is so clear that there is no factual dispute, these issues frequently involve the need for testimony and an evaluation of credibility by

---

[2] Federal Rule of Civil Procedure 26(b).

a fact finder. Although a grant of summary judgment may be difficult at times, it may be desirable as a means of framing the issues in an effort to expedite the proceedings.

17.5.6 Trial

The trial phase of litigation can be long and expensive in both time and money. The more issues and the more complex the issues are, the longer the trial is likely to take. Patent and trade secret cases can have lengthy trials due to the need to educate the fact finder on the technology at issue. Trademark and copyright trials can also be time consuming; however, some of these cases can involve simple issues that are easily comprehended by the fact finder.

Experts are especially important in intellectual property litigation, particularly in trade secret and patent litigation where complicated technological issues are likely to be present. An expert witness in a patent case may be needed to testify as a "person having ordinary skill in the art." An expert can also help by assisting the judge and jury in gaining an understanding of technology involved in the case. Experts are also frequently used in the economic area to establish or refute damages.

17.5.7 Costs

Some statutes make attorney's fees and costs available as remedies to intellectual property owners involved in litigation. The Patent Act and Trademark Act both make attorneys' fees available in "exceptional cases." The Copyright Act permits a court, in its discretion, to award litigation costs including attorney's fees to the prevailing party. Courts usually interpret each of the provisions as applying to cases where willful infringement has been found. U.S. courts do not routinely grant fee shifting awards and litigation should not be viewed as a "loser pays" situation.

17.6 Proceedings in the US Patent and Trademark Office

17.6.1 Trademark Oppositions

A trademark opposition is an *inter partes* (meaning "between the parties") proceeding that may be requested by any party who believes there is a potential to be damaged by the registration of a trademark. After a trademark application is examined and found allowable, the U.S. Patent and Trademark Office publishes the mark in its *Official Gazette* to put the public on notice of the potential registration of the mark. A potential opposer has 30 days from the date of publication to file an opposition, or request an extension of time for filing an opposition. Typical grounds for filing an opposition include likelihood of confusion with the opposer's mark, genericness or descriptiveness, abandonment, and fraud.

After an opposition is filed, the trademark applicant/owner responds by filing an answer asserting any defenses. Frequently, the answer to an opposition will assert a counterclaim to cancel a registered mark of the opposer which forms the basis of the opposition. The opposition proceeding is held before the Trademark Trial and Appeal Board. This is an administrative panel, rather than a judicial panel, with expertise in trademark matters. The proceeding is conducted similar to a federal court proceeding, but is less formal. The PTO proceeding concludes with a Board decision either refusing the registration or permitting the pending registration to go forward to registration. The losing party may appeal the Board's decision to the U.S. Court of Appeals for the Federal Circuit. The denial of a registration does not mean that the mark cannot be used. The right to use the mark requires a judicial determination; however, the results of an opposition can often lead to a settlement between the parties.

17.6.2 Patent Interference

For Patents Filed Before March 16, 2013

The Leahy-Smith America Invents Act effectively eliminated patent interference proceedings. However, patents that claim a filing date prior to March 16, 2013 will still be subject to interference proceedings as they are governed by the first-to-invent patenting system. A patent interference proceeding is a highly technical *inter partes* proceeding that is declared when two inventive entities are claiming the same patentable invention. In order to have interfering subject matter, there must be a reasonable possibility that the first entity to file a patent application was not the first to invent. This is a condition of establishing a right of priority to the inventive claims.

Priority of invention is decided in an interference proceeding conducted before the Board of Patent Appeals and Interferences, another administrative panel within the USPTO. The party who filed the earlier patent application is designated as the "senior party," and the later filer is known as the "junior party." For determining priority, the concept of invention is broken up into two parts: (1) conception of the claimed invention and (2) due diligence in the reduction to practice of the claimed invention. Conception is the mental element of invention. Reduction to practice happens subsequent to conception and it may be based on an actual reduction to practice or constructive reduction to practice. For an actual reduction to practice, the inventor must construct a working embodiment of the invention. Constructive reduction to practice results from the filing of a patent application.

In determining priority of invention, the junior party has the burden of proving an earlier date of invent. This can be done in two ways. First, the junior party can show an earlier date of conception coupled with diligent reduction to practice, either actually or constructively. The senior party's evidence may establish a conception date prior to the patent application date and a subsequent reduction to practice.

At the conclusion of the interference proceeding, the Board issues a decision awarding one party priority to the subject claims and the applications are returned to the examiner, who is bound by the Board's findings, for final examination.

For Patents Filed After March 16, 2013

As part of effort to harmonize the U.S. patent system with the patent systems of the rest of the world, the U.S. switched to a first-to-file system on March 16, 2013. A first-to-file system renders the question of who invented the claimed invention first moot. Thus, patent interference proceedings will no longer be available for patents that claim a filing date after March 16, 2013.

17.6.3 Patent Reexamination

A reexamination may be ordered at any time after a patent has been granted. Anyone, including the patent owner, may file a request for reexamination, which shows the existence of a substantial new question of patentability. Reexaminations are most frequently requested in two situations. First, a competitor or potential infringer seeks to have a patent invalidated and removed from the commercial arena. Second, the patent owner wants to enforce the patent, but has concerns about the patent's validity. Reexamination may resolve that issue, and it has the potential to strengthen the patent by bringing the additional prior art not applied during examination to the attention of the U.S. Patent and Trademark Office. Reexamination proceedings may be conducted either as an *ex parte* proceeding or as an *inter partes* proceeding.

Reference

1. American Intellectual Property Law Association, *AIPLA Report of the Economic Survey 2011*, American Intellectual Property Law Association (2011).

Chapter 18
Monetization Strategies for Startups, Incubators and Accelerators

18.1 Start-ups

Building a successful start-up is rewarding, and there are many challenges to be faced as the start-up moves towards becoming profitable. A start-up will transform and morph as it escalates through the different stages of growth, each having unique financial needs and financing options. The stages of development include seed, start-up, early-stage development, growth and maturity. Securing funding early on is one of the most challenging endeavors a start-up undertakes. It is understandable that a start-up with limited capital may view obtaining IP rights as waste of precious funds. This perception often leads start-up companies to the presumption that it is more affordable and a better business strategy for the start-up to utilize first-mover advantage to monetize newly developed technology, rather than bother with obtaining IP protection. While first-mover advantage has the benefit of turning the company's newest technological development into cash quickly, the risk associated with utilizing first-mover advantage as a business strategy is high and can doom an up-and-coming company once its competitors get a hold of its unprotected technology. If the start-up has developed something that triggers the development of a whole new field of hot technology, or serves as a base platform for the development of that new technology, the implications of being the owner of that base-level IP could be immensely profitable if the IP rights are properly protected.

Rather than viewing the cost of IP protection as a fund-draining burden on the start-up's limited capital, obtaining IP rights should be viewed as an investment in the start-up. If a start-up wants to become an established, successful business, the start-up needs to behave in the same way as an established, successful business—focusing meticulously on maximizing the return on every dollar spent.

The thought of pursuing a patent, for example, can be daunting for a small start-up. The cost of a typical provisional patent application ranges from $3,000 to $7,000. The cost of a typical non-provisional patent application ranges from $8,000 and $12,000. Then, if the patent issues, there is the cost of issuance, and the cost of maintaining the patent; i.e., maintenance fees to be paid to the USPTO at

G. B. Halt, Jr. et al., *Intellectual Property in Consumer Electronics, Software and Technology Startups*, DOI: 10.1007/978-1-4614-7912-3_18, © Springer Science+Business Media New York 2014

3.5, 7.5, and 11.5 years after issuance of the patent. There may also be the cost of litigation, if any is required, to enforce the patent. While these are all necessary costs incurred during the patenting process and during enforcement of the patent, a determined and savvy start-up company would be looking for a way to make the cost associated with procuring patent prosecution services more manageable.

In an effort to allay a start-up client's concerns regarding legal fees, some law firms offer start-up company clients special fixed-fee rates that are available exclusively for start-up clients. For example, when filing for a patent, the start-up client would pay a reasonable set amount for the patent filing and prosecution, regardless of how many hours the IP attorney actually spends on prosecuting the patent. This pricing model generates value for the start-up company client as patent prosecution can potentially require many back-and-forth communications between the patent prosecutor and the Examiner assigned to the patent application at the USPTO.

Additionally, some law firms will negotiate a payment structure with start-up company clients. A busy start-up, with limited funds now, needs a law firm that understands the company's financial situation. For instance, while there is an initial up-front cost for filing a patent (the cost of preparing the application for submission to the USPTO and payment of the filing fee), other costs associated with the patent can be paid at a later date as those costs arise, i.e., the issuance fee can be paid at the time the patent is allowed or maintenance fees, which are due at 3.5, 7.5, and 11.5 years after issuance of the patent, can be paid once they are due.

18.1.1 IP Portfolio Development and Investors

IP rights can serve as valuable company assets and can be monetized in several ways. For example, the addition of a patent, or simply a patent application, to a start-up's IP portfolio makes the start-up an eye-catching investment opportunity to equity financiers—venture capitalists (VCs) and angel investors who view a start-up's attempts to obtain patent protection as a demonstration of the company's dedication and commitment to striving for success. Pending patent applications can also be used by investors to assess whether the start-up will effectively be able to deter potential competition.

Owning IP, or even simply pending patent applications, can set your start-up apart from other, less confident start-up ventures. Some VCs and angel investors will not even consider investing in a start-up that lacks an IP portfolio, or at least basic IP for the company's core technology. Both large investment banks and boutique private equity firms are interested in raising and investing funds that are specifically targeted at those small entities in possession of IP rights and other intangible assets. While the initial cost of prosecuting a patent may seem expensive to a start-up, the return on investment for obtaining a patent, or other IP rights, could secure the start-up the investment capital it needs to launch into a profitable and successful enterprise.

18.1.2 Positioning a Startup to be an Attractive Acquisition Target

Some start-ups' primary business strategy is to utilize the contents of their IP portfolio as part of an exit strategy, with the goal of making the company an attractive acquisition target for larger corporations or competitors. Often times, it is more cost effective and a good business investment for a large corporation or competitor to buy-up a small start-up that holds desirable IP, rather than to license the IP from the start-up or risk infringement litigation. This can often result in a high rate of return for the start-up since the buyer is paying for the patent, as well as the potential future benefit of the patent. If the start-up is not looking to be acquired, but is seeking to monetize the patent, the start-up could utilize a sale and lease back acquisition model whereby the start-up sells the patent to a competitor, but requires in the terms of sale a covenant not to sue, which is essentially a license allowing the start-up to continue using the patented invention without the consequence of infringement. Effectively, the start-up would capitalize on the patent without losing the right to use it.

Factors that make a company an attractive acquisition target include:

- IP portfolio breadth and scope
- IP insurance
- Complete ownership of the IP in the portfolio
- The quantity of patent applications verses the number of issued patents
- The overall quality of the IP
- Comprehensive protection afforded by the IP in the portfolio.

18.1.3 Case Study: Facebook's Acquisition of Instagram, Face.com and Snaptu

Facebook has become a ubiquitous social networking service. Its popularity has grown immensely since its debut on the Internet in February of 2004. Having more than one billion active users as of September 2012, it is the fastest growing social networking site in terms of popularity. Facebook, as a company, has grown immeasurably and makes headlines every time it acquires a new business. Facebook has made a series of acquisitions over the years to supplement the technology behind many of the user functions available on Facebook's social media site. IP related to such supplemental technologies have added value for Facebook since the technology is related to the core of Facebook's social media business. Several examples of Facebook's strategic acquisitions include acquiring Instagram, Snaptu and Face.com.

The Acquisition of Instagram

Instagram, an instantly popular photo sharing service application launched in 2010, amassed approximately 30 million users by the time of its Android mobile platform release in April of 2012. Within the first day that the Instagram application was available for download on the Android platform, more than 1 million individual users downloaded the app. Facebook noticed Instagram's success and expressed an interest in acquiring Instagram shortly thereafter. A deal was negotiated in late April of 2012, consisting of an acqui-hire arrangement (the employees from the acquired company are part of the deal), $300 million in cash, plus 23 million shares of Facebook common stock in exchange for Instagram. The deal was not finalized until September 2012 when the Federal Trade Commission approved the deal. At the time of the negotiations in April, Facebook stock was valued at nearly $38 per share, but by the time the Federal Trade Commission had approved the deal, Facebook's stock price had dropped to around $19 per share; what started as a billion dollar deal for Instagram ended up being worth $715 million upon closing. During negotiations, Instagram was probably disinclined to negotiate for a floating share exchange ratio since Facebook's share prices were experiencing growth. But after Facebook's IPO in May 2012, Facebook's stock value dropped, and Instagram ultimately did not get the $1billion it initially set out to earn on the deal.

Facebook had consistently experienced difficulty with photo uploading on mobile platforms. A user would have to take a picture with the phone's camera functionality, and then upload the picture to Facebook by accessing the photo stored in the phone's photo library. Acquiring and implementing the technology behind Instagram would allow users to access an application which allows them to take photos and instantly upload those images to Facebook. In theory, the technology transfer from Instagram would enable Facebook to streamline the mobile photo uploading process that Facebook users are so fond of. The strategic value of the technology acquired by Facebook is readily apparent; however, by 2013 Facebook still had not effectively implemented the Instagram technology it acquired. Instagram was an acqui-hire acquisition and as such the Instagram team of software developers and engineers was absorbed by Facebook. The Instagram department has grown in number of employees since the acquisition, which has contributed to idea generation and the development of new products that Facebook is not using. Facebook has been criticized for its "acquire now, and figure out what to do with the acquisition later" approach to the Instagram acquisition.

Then in December of 2012, Instagram announced proposed changes in its terms of use, due to take effect in early 2013. The changes would give Instagram the right to share photos uploaded with the Instagram app to its parent company's site, Facebook. Similarly, language in the revised terms of service would also grant Instagram the right to sell users' photos to "business or other entity … without compensation to [the user]." Users cried out in outrage and expressed concerns about copyright ownership in the photos that were posted.

The Acquisition of Face.com

In June of 2012 Facebook acquired the small start-up company Face.com, primarily to access Face.com's facial recognition software IP rights. Face.com focused on the development of facial recognition software for use on mobile platforms. The software allows a user to "teach" the application the name of the person, or subject, in the picture. After learning the photo subject's name, the application can search through photos and automatically tag any other pictures of that same person with their name. Such technology would fit nicely with Facebook's social media site capabilities, as Facebook users tag photos of their friends and themselves constantly. Facebook's planned course of action involves coupling the Face.com technology to the technology obtained from the Instagram acquisition, making it easy for Facebook users to upload pictures and automatically tag those photos.

The Acquisition of Snaptu

Snaptu is a social media mobile application created by Mobilica (the company later changed its name to Snaptu), which is a small start-up company that creates java-based applications for mobile phones which replicate many of the features of smartphone applications, thereby making those same applications accessible on mobile devices that are not smartphones. The Snaptu technology is compatible with over 2,500 different mobile devices. Facebook acquired Snaptu in March of 2011 for an estimated $70 million. The deal was an acqui-hire situation; Facebook gained the Snaptu employees in addition to the technology and the related IP. Facebook's strategy was to acquire technology that would enable Facebook use on ordinary mobile phones so that the social media site could enter emerging markets that do not yet have access to smartphone technology.

18.1.4 Liquidating IP Assets to Recoup Losses

In the event that the start-up fails, patents can be monetized in a way to help recoup losses incurred by the failure of the business. There is a logical strategy in the start-up industry that if the start-up is going to fail, it should fail in the quickest, most efficient and cost-effective way possible. This strategy makes good business sense because the less cost incurred during failure translates to less net loss as a result of the failure. Patents can be bought and sold like assets. In the event that the start-up goes under, the start-up can sell its patents, or patent applications, as a means to recover some of the losses sustained by the company. The start-up can sell or auction off its IP portfolio, or portions of its portfolio.

18.2 Incubators

Business incubators are designed to nurture small, fledgling businesses during the start-up phase by providing business development support and encouraging the business to flourish and grow at its own pace. A cluster of quasi-related burgeoning businesses make up a business incubator. These businesses share a workspace where they can all operate harmoniously. Incubators provide the businesses with services and access to office equipment, telecommunication/Internet services and reception services. Incubators also provide participants with an extensive network of well-connected professionals, experienced mentors and an array of specialized experts in the relevant technology field.

In order to participate in an incubator, the small business must apply and be selected for participation in the incubator. Although incubators can vary greatly according to their specific goals or "mission", some application criteria may include:

- The breadth and depth of the start-up company's business plan
- The composition and quality of the members of the business's management team
- The likely capitalization capacity of the company's idea, and
- The estimated time to commercialization.

Typically applicants are selected based on whether the applicant business is compatible with the other businesses in the incubator and the space that is available for use. Other considerations are also taken into account, such as whether special permits are required for the applicant business to operate, whether specialized equipment and or laboratory space is needed, etc. Once admitted to the incubator, participant companies typically stay for a period of 1–3 years although this duration may be significantly more or less depending upon the incubators.

There are a variety of benefits available to the participants of a business incubator. For example, the incubator environment is highly conducive to networking and professional development. Participant businesses are provided with numerous opportunities to connect with other professionals, entrepreneurs and mentors. Inexperienced entrepreneurs can gain exposure to new ideas, garner tried-and-true business advice from successful serial entrepreneurs, and are encouraged to make connections and business contacts that they otherwise could never possibly establish.

Business incubators also offer educational programming for those businesses that are participating in the incubator. Entrepreneurship educational programming often includes seminars, workshops, lecture series and panel discussions on various topics including strategies for approaching funding sources, pitching ideas to investors, managing intellectual property portfolio, media relations, the importance of marketing, and how to manage growth of the company.

The reduced costs associated with participating in an incubator are substantial as well. The work space is offered to the business at a reduced rental rate. Furthermore, the incubator participants share overhead expenditures for the facility and split the cost of basic business operation essentials (electricity, Internet, utilities, etc.). Incubator participants also benefit from any volunteer initiatives instituted by the incubator. Many incubators have volunteer initiatives where professionals in the business community offer their time and expertise to incubator participants free of charge. Businesses within the incubator can receive targeted advice and professional services (such as legal, accounting or IT services) and the opportunity to make connections with these professionals as they work with the incubator participants to build a successful business.

18.2.1 Case Study: The University of Pennsylvania's UPSTART Company Formation Program

The UPStart program at the University of Pennsylvania in Philadelphia, offers support and guidance to start-ups that are trying to commercialize new technology. The program began in 2010 to address the need for commercialization support for technology developed by University research efforts conducted at the University of Pennsylvania. Implementing a business incubator within an educational institution helps to vertically channel new technology developed by the University through the commercialization process. The UPStart program puts these newly developed technology ventures into contact with other valuable resources, such as renowned faculty members and research entities, service providers, experienced entrepreneurs, potential investors and marketing opportunities. The Program has also established collaborative relationships with several local research and business entities, including the University Science Center and the Wharton Small Business Development Center to help get new small businesses up and running successfully.

Protocol at the University requires that faculty members who invent new technology disclose the invention to the University's Center for Technology Transfer. If the technology seems like a worthy investment, intellectual property rights are sought for the invention, and the technology is introduced into the UPStart program. A company is formed, and the company signs an agreement with the UPStart program, forging a partnership and obliging the company to adhere to UPStart's rules. The company must form an LLC, incorporating in Delaware, and must file all the requisite forms associated with doing business with the government. Once the company is established, the company and the UPStart program work together to develop a business plan and marketing strategy to attract funding from prospective investors. The program will also assist the company in writing grant proposals. Also, in an effort to educate the fledgling company and to help move the business along, the company will be assigned an experienced entrepreneur by the program. The company will work with outside legal counsel to create an IP management strategy specifically tailored to the unique technology and will explore licensing options. Once funding is secured, the IP rights accrued thus far are transferred to the company.

Additional perks available to UPStart participants include benefiting from the variety of services provided by the program. Participants also benefit from discounted rates offered on vendor provided services due to UPStart's business relationships with a variety of preferred venders. Furthermore, companies can lease laboratory and/or office space at a rate of $1/sq. ft./month.

18.3 Business Accelerators

Incubators and accelerators share similar characteristics, such as offering business development services and professional advice and guidance to start-up participants. Also, businesses that participate in an accelerator, like an incubator, have a myriad of opportunities to network with others in the field. The major difference between incubators and accelerators is time and the amount of pressure that the start-up is put under. Business accelerators are intense, time-compressed programs designed to assist start-ups in getting up and running in a matter of months.

A business accelerator is intended for start-ups that are focused on reducing their time to market. Application to an accelerator is often open to anyone. Once selected as a participant in an accelerator, the process moves very rapidly and is focused on team success, rather than individual development or mentoring. Accelerators are particular when choosing prospective participant companies. The primary focus of an accelerator is to provide participant companies with exposure to potential investors, and by the same token, provide investors with what can be considered quality, promising new start-ups worth investing in.

18.3.1 Case Study: The Y Combinator Business Accelerator

Y Combinator is an information technology start-up accelerator that was founded in 2005 by Paul Graham, Trevor Blackwell, Robert Morris, and Jessica Livingston. It provides start-ups with seed funding and invests approximately $18,000 per investment cycle into the participating start-ups of that cycle. In exchange for the seed funding provided by Y Combinator, Y Combinator will retain stock in the successful start-up; approximately 2–10 % but typically falling in the 6–7 % range.

After completing an application, selected applicants travel to the San Francisco Bay Area to attend the program during one of two three-month funding cycles: January-March or June–August. The accelerator operates in a similar fashion to a business incubator. Y Combinator offers participants all the necessary services and forms to turn the start-up into a fully operational business entity.

Y Combinator is designed to encourage collaboration between participants; participants demonstrate a prototype of their project within two weeks of starting the program and are regularly encouraged to openly talk about their projects with each other during the program in hopes that participants will help each other arrive at solutions and improve their ideas. During the session, participants attend numerous workshops and events designed to educate the participants and address problems that they are having with the commercialization, vision, and strategy or product development. Some of these sessions are geared toward developing a good pitch for the product, while others are intended to be counseling sessions with an investment firm so that participants can determine the best strategy for procuring funding based on the particular technology being commercialized. Participants can also schedule time to discuss these issues with any of the 9 members of the Y Combinator staff who have demonstrated experience and success in digital start-ups as they are either the founders of Y Combinator or they are the older alumni of the program. Participants also attend weekly dinners where they have the opportunity to share and demonstrate completed projects or progress while having the opportunity to listen to notable guest speakers during the meal. Notable guest speakers are generally prominent figures in the digital start-up space and have included successful individuals such as Al Gore, Alexis Ohanian (co-founder of social news website Reddit), Mark Zuckerberg (Facebook founder), Marissa Mayer (Yahoo president and CEO) and Rich Miner (co-founder of Android Inc.).

The session concludes with a Demonstration Day after 11 weeks of highly productive time spent developing the start-up. Participants demo a final product to investors and interested spectators. This is an opportunity to attract funding and also an opportunity to practice all the skills learned during participation in the Y Combinator accelerator program. Some investors are so captivated by what they see that they indicate their commitment to funding the idea at Demo Day, while others require additional convincing in subsequent meetings.

Y Combinator has funded more than 460 start-ups since its inception in 2005. Some of the most well known companies to develop as a result of Y Combinator funding include Dropbox (cloud storage space where users can store files) and Scribd (social reading and publishing website). When the company has become sufficiently successful, it exits Y Combinator either through acquisition or IPO. A few famous exits from Y Combinator include Reddit (social news aggregator website), Couldkicker (centralized web-based control hub for cloud servers), and OMGPOP (game developer/publisher of multiplayer, social and mobile games). Despite having exited, alumni companies can continue to access the network of alumni after the program has ended.

18.4 Typical Services Offered by Business Incubators and Accelerators

Incubators and accelerators offer participant businesses a collection of services to promote the development and growth of the business participants. They may provide educational programming, business plan development workshops, customer service training, mentoring, and networking opportunities. Business incubator managers act as liaisons and facilitators, putting the entrepreneurs in connection with firms, specialists and experts to render unique services, consultation and advice. Incubators and accelerators also make an effort to assist participants in gaining access to seed capital and federal grants. They offer services to help businesses prepare financial proposals and grant applications, assist participants in obtaining loans and introduce participants to Angel and venture capital investor networks.

Some incubators and accelerators limit the type of businesses which may participate so that the participants can share research equipment. For example, the businesses may be restricted to a certain industrial field such as biotechnology or life sciences. These particular businesses require the use of wet laboratory space, specialized lab equipment, permits for dangerous chemicals or other hazardous materials and sophisticated scientific instruments. To cut down on costs, the businesses within the incubator or accelerator share a communal lab space, with designated time slots when the businesses may use the lab area.

Chapter 19
Monetization Strategies for Universities and Research Centers

For universities and research entities, translational research is both a source of revenue generation as well as a source of academic notoriety. The ultimate downstream purpose of translational research is two fold. First and foremost is that translational research is intended to solve real-world problems via practical application of newly developed or innovative technology. Second, translational research is meant to have commercial value and to generate unrestricted income for the university; either through the development of a marketable product, or through licensing agreements. Monetization of translational research efforts can be highly lucrative. There are a wide variety of ways to generate revenue from research. Furthermore, there are strategic and synergistic business relationships that can be formed between a university or research entity and industry partners or corporate sponsors.

Table 19.1 shows the top five research entities and their respective licensing revenue from translational research efforts for the 2010 fiscal year.

19.1 Bayh-Dole Act: Commercialization of Federally Funded Innovation Efforts

Universities and research entities are often the recipients of federal funding in the form of government research grants or financial incentive programs. The University and Small Business Patent Procedures Act,[1] commonly known as the Bayh-Dole Act, enables small businesses and non-profit organizations to retain their IP rights to innovations resulting from federally funded research projects so long as the research entity complies with a set of predetermined rules. For example, fund recipients must share royalties with the inventor and actively promote and attempt to commercialize inventions that result from the research funding.

The Bayh-Dole Act also gives the government the power to compel licensing of technology developed from public funding on reasonable terms, yet it has never

---

[1] 35 U.S.C. § 200–212.

G. B. Halt, Jr. et al., *Intellectual Property in Consumer Electronics, Software and Technology Startups*, DOI: 10.1007/978-1-4614-7912-3_19, © Springer Science+Business Media New York 2014

Table 19.1 Data from the Association of University Technology Managers. Numbers are for the 2010 fiscal year

Research entity	Licensing income
Beckman Research Institute of City of Hope National Medical Center	$202,300,000
Northwestern University	$179,800,000
New York University	$178,400,000
Columbia University	$147,300,000
Memorial Sloan-Kettering Cancer Center	$139,400,000

exercised this power. There are two situations in which the government may compel licensing under the Bayh-Dole Act. First, the government may require a federally funded patentee to grant a reasonable license to a responsible applicant by exercising its "march-in rights" in situations where government intervention is necessary to prevent the anticompetitive nonuse of patents. Second, the government may compel a royalty-free license to practice any patented technology which has been funded by the government.

Federal research funds are designated for a specific purpose; i.e., a research grant to investigate electrical impulses of the human heart during sleep must be used explicitly for that purpose. However, if the researchers were to invent a new electrical impulse measuring device during the course of their research, any revenue resulting from commercialization efforts of the measuring device could be used for whatever purpose the university sees fit—it is unrestricted income for the university. The result is that universities are encouraged to collaborate with business entities in industry to promote the application of their inventions.

19.2 Utilizing the Technology Transfer Office

Most universities and research entities have a technology transfer department or office. A tech transfer office's primary function is to generate revenue from the translational research efforts of the university or research entity. Each technology transfer department at each university or research entity is unique by necessity. The research initiatives differ at each location, the need to generate revenue from technology transfer varies, and the methods and monetization strategies employed by universities and research entities differ dramatically. For example, technology transfer offices are involved in negotiating corporate-sponsored research agreements, marketing and commercialization of inventions, managing and monitoring existing license agreements for compliance with licensing terms and protecting against potential abuses, evaluating and analyzing new invention disclosures, overseeing patent procurement, maintenance and licensing, and coordinating with the public relations department concerning announcements of new innovations developed at the university or research center.

Tech transfer offices can also collaborate to manage multi-university patent pools. Similar to how any other patent pool is formed, multiple universities collaborate by contributing unlicensed, technologically-related patents into a common collective. A company seeking access to the pool would sign a single licensing agreement in exchange for the right to practice all the patents in the pool. In order for a pool to be successful, the university contributors must determine a licensing royalty payment structure that is suitable for all contributors to the pool. Patent pooling among universities is often appealing to corporations. Not only does the corporation gain access to the wealth of IP available in the pool, but there are also valuable strategic alliances to be made with the various university pool contributors.

19.3 Case Study: University of California's Technology Transfer Program

The Technology Transfer System of the University of California is an exemplary model of a streamline process that takes university research initiatives and translates them into useful products for commerce. With 10 campuses making up the University of California, the University is constantly pumping out new inventions in need of patent protection. The University has an active patent and patent licensing program under which employees of the University, and anyone who uses University funding, facilities or supplies in conducting research, agree to disclose any patentable invention to the University so that the University may seek patent protection. The inventor assigns all rights in the invention to the University, but the University agrees to share royalties earned on licensing agreements with the inventor. Payouts to inventors are made in November and are based on royalties generated from the previous fiscal year (ending in September). The University's policy also includes a provision related to inventions that result from sponsor funded research. For sponsor funded research, the rights are generally assigned to the sponsor.

All disclosed inventions are assessed by the University's IP counsel for patentability, commercial viability, pending publication bar dates and licensing potential. A prior art search is performed and a determination is made as to whether a patent should be filed. For some technology, international patent rights are also sought. In the event that the University does not opt to pursue a patent on the invention, the University may elect to release the rights to the inventor.

Once a patent application has been filed, the University contacts potential licensees and assesses their level of interest in the IP rights to the invention. Once a prospective licensee is identified, the parties negotiate the terms of the license. Licensing negotiations are individualized and specifically tailored by the parties involved in the licensing agreement. The parties will determine terms like duration of the licensing agreement, royalty rate and the scope of the license. Conversely, if the research resulted from federal funding, then, in accordance with the Bayh-Dole Act, the federal government is given a royalty-free license by the University.

Technology that is available for non-exclusive licensing is made available to the public in a listing found on the University's website. A brief description of the technology is provided, along with the patent status of the technology and contact information. Generally, the licensee will pay a licensing fee and will provide the University with the negotiated royalty on all sales related to the invention. Furthermore, the licensee will often be responsible for reimbursing the University the costs associated with patent procurement. The licensee is charged with commercializing the invention under license.

19.4   University and Research Entity Partnerships

Universities frequently partner with research entities, such as research hospitals, research centers and other universities. A university can derive funding or income from these partnerships in several ways. For example, some grants require a partnership between two research entities as application criteria for the grant funding. There is also the likelihood that new IP will be developed as a result of the combined research efforts of the two entities. Furthermore, joint research efforts between universities and research entities are often sponsored by corporations.

Example: In 2010, St. Jude Children's Research Hospital partnered with Washington University in an effort to identify the genetic causes that result in some of the most deadly childhood cancers. Some of the cancers to be researched included acute megakaryoblastic leukemia (a rare form of leukemia), aggressive eye tumors and common medulloblastoma (brain tumor). The initiative also sought to establish a genome project for pediatric cancers.

19.5   University and Incubator/Accelerator Partnerships

A university or research entity can exploit IP rights through the creation of a new company; either a spin-off or a start-up. Typically, the university or research entity will retain a shareholder interest in the newly-formed company. To successfully transition from a research initiative to either a startup or spin-off venture requires a well thought out business plan, a strong and respectful management team, relentless motivation, unremitting determination and support. Universities often create these new companies through involvement with business incubators or accelerator. The university could either operate the incubator itself, or could partner with an existing business incubator. For more on the benefits of involvement with business incubators and accelerators, see Chap. 18: Monetization Strategies for Start-ups and Utilization of Incubators and Accelerators.

19.5.1 University Operated Business Incubators and Accelerators

A university owned and operated business incubator or accelerator promotes the development of innovative research as well as educational programming. A university operated incubator can be used for internal purposes and can also generate revenue for the university by accepting participants from outside of the university community. By accepting business proposals from ventures outside of the university, synergistic business relationships are formed and the incubator can serve as a development and resource center for new start-up and spin-off companies while also providing an experiential learning environment for students at the university.

Incubators and accelerators are designed to nurture small, fledgling businesses as they grow into flourishing successes. The incubator/accelerator offers rental space and business development services to those start-up and spin-off companies participating in the incubator/accelerator. Universities generate income from these activities and also derive a cost savings when using incubator participants as subcontractors for various other research initiatives.

19.5.2 Use of a Business Incubator or Accelerator Outside of the University

Independently operated incubators are often the result of collaboration among multiple innovative enterprises focused on promoting entrepreneurial endeavor, technological advancement and economic development. Partnerships between a university or research entity and an independently operated business incubator or accelerator have many benefits, including the formation of synergistic relationships between the university and the incubator enterprises, access to a wealth of experienced and successful entrepreneurs and other business participants of the incubator and cost savings associated with involvement in an incubator, such as reduced rent and overhead cost expenditure and free or reduced rate advice and professional services.

19.6 Universities and Corporate Sponsorship

Corporation sponsored university research is an ideal business solution in situations where a company wants to develop specific areas of its business. This university-business partnership often results in a synergistic business relationship between the two parties since both parties share a vested interest in the outcome of their combined efforts. The interactions between the company and the university can include: in-licensing, sponsored research, collaborative research, faculty consulting, sharing of research materials, joint ventures or spin-offs, and donations of IP.

19.6.1 In-Licensing and Out-Licensing

In certain situations, a university may perform research or develop a technology that a corporation has a vested interest in gaining access to. The company could engage in a form of technology transfer, or licensing, known as in-licensing, where the company pays the university a fee in exchange for access to certain IP owned by the university. Out-licensing is the opposite arrangement: the company would offer its IP to others (the university, for example) in exchange for a fee.

19.6.2 Sponsored Research

Corporate sponsorship of university research can be a mutually beneficial endeavor. Corporate sponsors directly benefit from the inventive talent pool and wealth of knowledge available exclusively in university laboratories. The sponsor typically receives basic research services at reduced costs as there is often a net lower development cost of the technology than if the research were performed in-house: human capital is more affordable and scientific equipment and other resources are often readily available in a university research setting. Cost-savings at the development stage of the product will translate into higher net profits once the product is brought to market. There is also a tax incentive, in the form of tax credits, for the corporate sponsor for investing in university research.

The benefits to the university are equally beneficial. First, the university receives much-needed research funding. Secondly, there is also a reduction in the cost of advancing technology, which allows the university to achieve more for each research dollar spent. Since the university programs achieve research advances with lower development costs, other potential sponsors from industry are encouraged to also grant sponsorship. Finally, some research is immensely capital intensive, and could not be explored without corporate sponsorship. Major companies in the pharmaceutical industry, oil and gas industries and automobile manufacturing have sponsored research at universities.

In corporate sponsorship situations, there are several technology transfer agreement options.

1. The university may assign the invention to the sponsoring corporation: all rights, title, and interests.
2. The university may agree to grant to the corporation an exclusive license for a designated period of time to develop and exploit the invention, but retain title to the IP rights to the technology.
3. The university may grant non-exclusive licenses to multiple corporations allowing the corporations to utilize the technology.

Example: The Semiconductor Research Corporation's Education Alliance sponsors an undergraduate research opportunity program at the University of Michigan, where students can complete a fellowship performing research related to the semiconductor computer industry. Students can participate in research projects relating to various aspects of semiconductors, including materials science engineering, electrical and chemical properties, or physics. Students work under the supervision and mentoring of University of Michigan faculty and gain access to Semiconductor Research Corporation's network of internship opportunities with major semiconductor companies, including Intel and Brookhaven Laboratories.

19.6.3 IP Right Donation

The donation of IP rights is an increasingly popular form of corporate philanthropy. Companies invest significant amounts of capital into developing innovations; but protecting those intangible assets must be balanced against cost considerations. Is the technology a core asset, or a non-core asset? Sometimes the R&D department develops a "good idea" that it ultimately turns tangential in nature, becoming an "orphan" technology. The technology may fail to mature to fruition, or may become misaligned with the company's business strategy. Incidentally, these "orphans," and any IP that has been acquired, become non-core assets, effectively sitting on the shelf, inactive and unused. Rather than lose their economic benefit through abandonment of the technology or its IP, a company can make a charitable donation of the IP to a university or research entity. Such donations serve two purposes for the corporation: (1) The donation drums up good public relations attention when the donation is disclosed to the public, and (2) the donors can take a tax deduction for the fair market value of the asset (the price the property would sell for on the open market).

A donation must also be potentially lucrative for the donee as well—there must be some potential economic viability to the IP. Corporations cannot simply dump unused and inactive patents on a university, calling it a donation, when there is no economic value in the IP. It is also preferred that the donated patents be relevant to the charter of the university to which they are donated. The result is that the university gains rights to practice new technology that is likely complementary to its current research efforts.

Donations can be made to a variety of worthy recipients. Other candidates for receiving donations include technology transfer centers at state, regional and national levels, think tanks, and non-profit research organizations. The decision to donate can be spurred by a multitude of reasons. The donation could be the byproduct of an annual company reassessment and preening of its IP portfolio or a merger or acquisition could have occurred resulting in a new corporate mission, rendering once useful technology as no longer relevant to the company's pursuits. Companies such as Ford, Dow Chemical and Proctor & Gamble have made donations of IP rights to universities.

Chapter 20
Non-Practicing Entities

20.1 What are Non-Practicing Entities?

There are a myriad of definitions as to what exactly a non-practicing entity (NPE) is, and there are many opinions as to whether they are a benefit to the patent system or a detriment. However, a generally accepted definition of an NPE is an entity that earns or plans to earn a majority of its revenue from the licensing or enforcement of its patents.[1] The patents typically belong to a single technological area, or a grouping of related technologies. The NPE does not practice the patents, meaning that the NPE does not produce any goods or provide any services based on the patents rights that are held. Rather, an NPE collects patents in order to assert those patents in litigation for infringement or to entract licensing agreements. Sometimes this behavior is described as "opportunistic" or "aggressive." However, a patent grants a right to exclude others from making, using or selling the patented invention. The grant of a patent does not obligate the patent owner to make, use or sell the patented invention. NPEs exercise patent enforcement within the confines of the law. For NPEs, litigation, in addition to negotiating licensing agreements with potential infringers, is the NPE's primary means of generating revenue on the patents it owns.

20.2 Support for Non-Practicing Entities

In the last five years NPEs have been playing a larger role in the patent system. By buying patent rights in the first place, NPEs provide inventors with compensation for their hard work and at the same time provide inventors with financing so that they can continue to do more of what they do best—invent. Furthermore, NPEs give inventors the opportunity to have their patent rights enforced in situations where inventors might not otherwise be able to enforce those rights themselves due to a lack of funding for litigation.

---

[1] See www.PatentFreedom.com.

G. B. Halt, Jr. et al., *Intellectual Property in Consumer Electronics,*
Software and Technology Startups, DOI: 10.1007/978-1-4614-7912-3_20,
© Springer Science+Business Media New York 2014

Patents derive worth from the fact that patent rights create a temporary monopoly for the patent holder and generate value in one of two main ways. The first is through sale or licensing of the patent rights. The second is by enforcement of patent rights. If a patent is infringed, the patent holder may assert the patent rights against the infringer. The infringer then owes the patent holder compensation for infringement of the patent holder's rights. As previously mentioned, a patent only grants an exclusionary right—a patentee may exclude others from making, using or selling the patented invention in the United States. A patent does not confirm a positive right, nor does it obligate the patent holder to make, use or sell the patented invention. All that is required in exchange for the patent right is a disclosure of the invention to the public. A patent holder is not required to practice the invention. Along analogous lines, an author isn't required to publish a copyrighted work, nor is an artist required to place the copyrighted painting on display.

Some patents are issued with a narrow scope of protection, meaning that in order to infringe the patent what the infringer must be very precise, as defined by the claims of the patent. On the other hand, patents can also be issued with very broad claims, giving the patent a very wide scope of protection. The broader the scope of the patent, the easier it is to assert the patent against infringers and prevail. Additionally, a broad patent can be asserted against a wider swath of potential infringers than a patent that is narrow in scope. NPEs specifically look to purchase patents that cover broad ideas or concepts because the pool of potential infringers is larger. This strategy makes sense from a business point of view. The NPE business model centers on acquiring property rights as an investment and extracting revenues from the property. In theory, NPEs are no different from real estate developers or antique dealers. Standing by while infringement occurs is in effect a waste of the patent right to exclude others. The patent system allows for what would seem like the hording of patents for the purpose of enforcing them.

More recently have been describing NPEs by using the pejorative term "patent troll." NPEs have been prevalent for quite some time. Consider that universities often hold patents derived from research initiatives, but often the universities do not commercially practice the patents they hold themselves, save for universities with business incubators. Universities generate revenue from patent rights through technology transfer programs that licensing the technology to companies. Technically, this makes universities NPEs. Similarly, commercial enterprises often create patent holding companies within their family of businesses specifically for the purpose of managing IP rights and portfolios. It is not uncommon for companies to amass huge troves of patents to improve their defensive IP position in the event of litigation. For example, in 2011 Google bought a trove of defensive patents from IBM to protect against suit from Apple or Microsoft regarding the Android operating system. Another example, also from 2011, can be found in the Rockstar consortium's purchase of Nortel's patent portfolio when Nortel entered into bankruptcy, (the Rockstar consortium consisted of six companies, bidding together: Apple, Research-in-Motion, Sony, Ericsson, Microsoft and EMC).

Entering litigation with an NPE is similar to entering litigation with any other party. Patents held by NPEs can be designed around or reexamined by the USPTO.

They can be held invalid in court and preliminary injunctions can be sought for them. When an NPE wins a major infringement suit, the media reports on the litigation as if it is a David and Goliath type-story except David doesn't deserve to win. However, when an NPE loses a suit it rarely makes headlines. NPEs win sometimes, lose sometimes, and from time to time enter into settlements, just like practicing entities do during litigation. In practical terms, there is little difference between the two.

20.3   Abuse of the Patent System by Non-Practicing Entities

It is difficult to make generalizations about NPEs since there are many different types of NPEs. It has been contended that some NPEs abuse the patent system. There have been instances where an NPE has taken an overly aggressive approach to enforcing patent rights, or has been accused of creating shell corporations for the purpose of establishing jurisdiction where a favorable outcome from litigation is likely, engaging in offensive and defensive patent aggregation, and tailoring its patent portfolio to specifically target a company for infringement litigation purposes. When these types of accusations are true, it casts NPEs in a bad light. Furthermore, during litigation, NPEs are often difficult entities to assert a counterclaim against as they do not produce products or services.

NPEs are often investment vehicles and as such seek to achieve a return on the investment in their patents. Some NPEs that do not develop their own technology or IP and instead purchase patents from others. Such NPEs are arguably not contributing to the innovative process that the patent. System was designed to promote. Typically this type of NPE does not develop the inventions covered by their patents into products or commercialize them in any way. In fact, it is often argued that NPEs regular practice of merely enforcing patent rights and seeking licensing agreements, sometimes at excessive and unreasonable rates, actually stifles innovation because companies are afraid of being sued, or are forced into a compulsory licensing agreement with the NPE.

NPEs, compared to those entities that practice the patents they hold, do not incur production costs associated with product development. So when an NPE draws a company into litigation, that company must pull resources away from production and product development, and divert it into the litigation. When an NPE files a lawsuit against a small company, the small company often does not have the resources available to fight the NPE in court and will be forced into a settlement agreement. Elimination of competition in the market place increases costs and reduces the number of technologies that consumers have available to them.

Redesigning products to get around patent rights or monitoring the IP landscape around a certain technology for new IP to arise can be costly and time consuming, and for many small entities or individual inventors such monitoring is simply not feasible. Because smaller companies are ill-equipped to cope with NPEs, often NPEs pressure smaller companies into settlement agreements under the threat of impending litigation. Acquiring ownership in an aggregation of related patents belonging to a single technology area often has enhanced litigation value for the patent holder since the various patents cover a wider scope of the technology. NPEs aggregate patents by purchasing small portfolios from small businesses, startups, or individual inventors. The aggregated patents are then used in litigation against larger entities, to obtain favorable jury verdicts.

NPEs are unique in that they can carefully craft an IP strategy, picking and choosing patents that are complementary for aggregation. The potential downside to the gamble of litigation for an NPE is relatively low. They often have very few physical assets, make no products and have no customers. If a patent asserted by the NPE is ultimately held invalid or not infringed, all the NPE has lost is the sunk cost associated with that patent and the litigation. NPEs do not have any motivation to back down, or even settle, unless settling on a licensing agreement is the whole purpose behind bringing the suit in the first place.

20.4 Offensive and Defensive Patent Aggregation

NPEs often engage in patent aggregation, however, aggregation is not limited to NPEs—companies also engage in strategic patent aggregation. There are two patent aggregation strategies: offensive patent aggregation and defensive patent aggregation.

Offensive Patent Aggregation

Whether it is sports, war, or IP law, there is nothing like having a strong offensive position. This type of patent aggregation involves purchasing, or amassing as it sometimes seems, patents in order to assert them against competitors and to extract licensing agreements from those competitors. This can be perceived as building a war chest of patents in preparation of litigation.

Defensive Patent Aggregation

A strong defensive position is also beneficial when protecting an entity's IP rights. Defensive patent aggregation involves purchasing IP rights to keep the valuable patent rights out of the competitors' reach. Simply put, if you own the patents, your competitors cannot. A good defensive strategy often centers on a technology area and the NPE or company buys up patents related to that technology area, effectively creating a monopoly on patents in that technology.

Appendix A

Confidential Disclosure Agreement

This Agreement made by and between CONDUCTWELL, having an address of _____, and BIG-ELECTRONICS, having an address of_____.

WITNESSETH THAT:

WHEREAS, CONDUCTWELL has developed a new semiconductor product, (hereinafter referred to as "the invention"), and is in possession of certain related confidential and proprietary information (hereinafter referred to as "proprietary information");

WHEREAS, CONDUCTWELL is interested in disclosing the invention and proprietary information to BIG-ELECTRONICS, in confidence, for further evaluation for product production and distribution rights; and

WHEREAS, BIG-ELECTRONICS is interested in receiving such information, in confidence, for conducting such further evaluation.

NOW, THEREFORE, for and in consideration of the foregoing premises, and of the mutual promises set forth below and for other good and valuable consideration, the receipt and sufficiency of which is hereby acknowledged, the parties, intending to be legally bound, hereby agree as follows:

1. CONDUCTWELL agrees to disclose to BIG-ELECTRONICS in confidence proprietary information relating to the invention.
2. BIG-ELECTRONICS agrees to accept and hold in confidence any and all proprietary information disclosed by CONDUCTWELL under this Agreement, except:

    (a) information which at the time of disclosure can be shown to have been in the general public knowledge;
    (b) information which, after disclosure, becomes part of the public knowledge by publication or otherwise, except through breach of this Agreement by BIG-ELECTRONICS;

G. B. Halt, Jr. et al., *Intellectual Property in Consumer Electronics,*
Software and Technology Startups, DOI: 10.1007/978-1-4614-7912-3,
© Springer Science+Business Media New York 2014

(c) information which BIG-ELECTRONICS can establish by competent proof was in its possession at the time of disclosure by CONDUCTWELL and was not acquired, directly or indirectly, from CONDUCTWELL; and

(d) information which BIG-ELECTRONICS receives without restriction from a third party, provided that such information was not obtained by said third party, directly or indirectly from CONDUCTWELL.

3. BIG-ELECTRONICS agrees that the proprietary information received from CONDUCTWELL shall not be used by BIG-ELECTRONICS, other than for evaluation and consideration for use as noted above and as otherwise agreed between CONDUCTWELL and BIG-ELECTRONICS.

4. This Agreement shall not be construed as granting any license or any other rights to BIG-ELECTRONICS.

5. BIG-ELECTRONICS agrees to restrict access to the proprietary information to those employees, agents and representatives who are engaged in that actual evaluation of the invention on a need-to-know basis and will require all such employees, agents and representatives to agree to maintain the proprietary information in confidence.

6. BIG-ELECTRONICS agrees that any improvements or modifications developed by BIG-ELECTRONICS in connection with the invention shall belong to CONDUCTWELL, and BIG-ELECTRONICS shall assign all rights in any such improvements or modifications to CONDUCTWELL.

7. Upon completion of its evaluation, BIG-ELECTRONICS agrees to return to CONDUCTWELL all information concerning the invention, including all photographs, diagrams, drawings, descriptions, prototypes, and notes, and any copies thereof.

This Agreement shall be binding upon and shall inure to the benefit of and be enforceable by and against the respective heirs, legal representatives, successors, assigns, subsidiaries, and affiliated or controlled companies of the parties hereto.

This Agreement shall be construed, interpreted and applied in accordance with the law of the State of _____. With respect to the subject matter of this Confidential Disclosure Agreement, the foregoing constitutes the entire and only understanding between the parties, and this Confidential Disclosure Agreement supersedes any prior or collateral agreements or understandings between the parties with respect to confidentiality.

IN WITNESS WHEREOF, the parties hereto have caused this Agreement to be executed as of the last date written below.

BIG-ELECTRONICS

Date:_____ By:_____

          Name:
          Title:

CONDUCTWELL

Date:_____ By:_____

          Name:
          Title:

Appendix B
Assignment

Dr. Richard Diode, residing at_____, a citizen of the United States (hereafter the undersigned), is the inventor of _____ for which the undersigned executed an application for United States Letters Patent, U.S. Patent Application No._____, filed, 201__.

The undersigned hereby authorizes assignee or assignee's representative to insert the Application Number and the filing date of this application if they are unknown at the time of execution of this assignment.

ConductWell, Inc., a Delaware Corporation, having a place of business at _____, (hereafter referred to as the assignee), is desirous of acquiring the entire right, title and interest in said invention, all applications for and all letters patent issued on said invention.

For good and valuable consideration, the receipt and sufficiency of which is acknowledged, the undersigned, intending to be legally bound, does hereby sell, assign and transfer to the assignee and assignee's successors, assigns and legal representatives the entire right, title and interest in said invention and all patent applications thereon, including, but not limited to, the application for United States Letters Patent entitled as above, and all divisions and continuations thereof, and in all letters patent, including all reissues and reexaminations thereof, throughout the world, including the right to claim priority under the Paris Convention or other treaty.

It is agreed that the undersigned shall be legally bound, upon request of the assignee, to supply all information and evidence relating to the making and practice of said invention, to testify in any legal proceeding relating thereto, to execute all instruments proper to patent the invention throughout the world for the benefit of the assignee, and to execute all instruments proper to carry out the intent of this instrument.

The undersigned warrants that the rights and property herein conveyed are free and clear of any encumbrance.

G. B. Halt, Jr. et al., *Intellectual Property in Consumer Electronics,*
Software and Technology Startups, DOI: 10.1007/978-1-4614-7912-3,
© Springer Science+Business Media New York 2014

EXECUTED under seal on this _____ day of, 201__ at

_____.

(Place)

Witness:

_____   _____(L.S.)

Dr. Richard Diode

State of

SS.

County of

On this _____ day of _____, 201__ before me personally appeared Dr. Richard Diode, to me known to be the person described herein and who executed the foregoing instrument, and acknowledged that he executed the same knowingly and willingly and for the purposes therein contained.

Witness my hand and Notarial seal the day and year immediately above written.

_____

Notary Public

My Commission Expires:

Appendix C
Preliminary Invention Disclosure

It is important to have complete, accurate information about the inventor(s) and the invention. If you have any doubts about any information, include it. Every applicant for a patent is under a duty to advise the U.S. Patent and Trademark Office about all known prior art. Use additional sheets of paper if necessary. If you have any questions, please get in contact with us. Mail and/or fax this form to:

Volpe and Koenig, P.C.
Counselors at Law
United Plaza, Suite 1800
30 South 17th Street
Philadelphia, PA 19103

Telephone: (215) 568-6400   Fax: (215) 568-6499

(1) Title of Invention:

(2) Description of the Invention:

(3) First written description of the invention was made by:
Date:

(4) First drawing of the invention was made by:
Drawing No.:   Date:

(5) First model was made by:
Date:

(6) Invention was first disclosed to:
Date:

G. B. Halt, Jr. et al., *Intellectual Property in Consumer Electronics,*
Software and Technology Startups, DOI: 10.1007/978-1-4614-7912-3,
© Springer Science+Business Media New York 2014

(7) First effort to sell the invention:
   Date:

(8) First use of invention:
   Date:

(9) Description of circumstances(s) and date(s) of any disclosure(s) that are not described above:

(10) If you know of other products, processes or machines like yours, describe them and how your invention differs from them.

(11) Identify any known patents and/or other publications which relates to your invention. (Attach copies, if available.)

(12) In the case of a process invention has a product ever been made by the new process? Yes____ No____. If yes, has the product ever been offered for sale, sold, or used? Yes____ No____. If yes, please describe the circumstances with dates.

(13) If anyone other than a named inventor had anything to do with this invention, state the person(s) full name, company affiliation, address and title and describe that involvement.

(14) If this invention is assigned or is to be assigned, identify the assignee.

Inventor:_____   Inventor:_____
(Print full name)   (Print full name)

Citzenship:_____   Citzenship:_____

Home Address:_____   Home Address:_____

       _____   _____

       _____   _____

Telephone No.:( )   -   Telephone No.:( )   -

Social Security No.: _____   Social Security No.: _____

Signature_____   Signature_____

Date_____   Date_____

Attachments to this form:

Appendix D
Draft License Agreement

This Agreement made this _____ day of _____, 2013 by and between ConductWell, Inc., a Pennsylvania corporation, having a place of business at _____ (hereinafter "CONDUCTWELL") and Insulate Technologies, Inc., a Delaware corporation, having a place of business at _____ (hereinafter "INSULATE").

WHEREAS, CONDUCTWELL has developed a product (hereinafter "the CONDUCTWELL Product") intended for distribution and sale in a nationwide market and has expertise in producing such product; and

WHEREAS, INSULATE has a nationwide distribution network and is a manufacturer and distributor of similar types of products and is desirous of obtaining the CONDUCTWELL product and know-how as well as the right to manufacture, distribute and sell the CONDUCTWELL Product;

NOW, THEREFORE, in consideration of the above and of the mutual covenants and obligations contained herein, and intending to be legally bound, the parties hereto agree as follows:

1. DEFINITIONS

1.1. "Licensed Goods" shall refer to CONDUCTWELL's Semiconductor product as described in CONDUCTWELL's U.S. Patent Application No. XX/XXX,XXX.

1.2. "Licensed Territory" shall mean North and South America.

1.3. "Affiliate", with respect to either party, shall mean any person or entity who controls, is controlled by, or is under common control with a party to this agreement, and includes, but is not limited to, parent corporations, subsidiaries, and sister corporations.

1.4. "Net Selling Price" is the gross selling price (*i.e.*, the dollar amount) actually paid to and collected by INSULATE in a bonafide arms-length transaction consummated or intended to be consummated by transferring title in a Licensed Good, less:

G. B. Halt, Jr. et al., *Intellectual Property in Consumer Electronics,*
Software and Technology Startups, DOI: 10.1007/978-1-4614-7912-3,
© Springer Science+Business Media New York 2014

(a) returns actually credited;

(b) actual losses experienced by INSULATE as a result of credits issued for such things as expired shelf life; and

(c) shipping charges separately charged to transferee; however, no deduction shall be made for any other costs incurred, such as, but not limited to, costs of manufacture, sales, distribution, or exploitation of the Licensed Goods.

2. LICENSE

2.1. CONDUCTWELL hereby grants INSULATE and Affiliates a sole license for the manufacture, distribution and sale of the Licensed Goods in the Territory. CONDUCTWELL will provide technological support and know-how to manufacture the Licensed Goods to INSULATE.

3. ROYALTY

3.1. INSULATE agrees to pay a royalty to CONDUCTWELL of ____ percent (_%) on the Net Selling Price of the Licensed Goods. Royalties shall be payable on sales made during the period starting on the date of this agreement and ending on January 1, 2014, and for each annual period following therefrom, during the term or extended term of this agreement.

3.2. Royalty payments made under this agreement shall be made within sixty (60) days after the end of each quarterly period during the term of this agreement.

3.3. INSULATE agrees to submit to CONDUCTWELL within sixty (60) days after the end of each quarterly period a written royalty report setting forth the amount of royalties due for the preceding quarterly period and the manner in which INSULATE calculated said royalties. INSULATE agrees to keep complete records covering all royalty bearing activities specified in this agreement in sufficient detail under its current accounting to enable the royalties payable hereunder to be determined and verified.

4. AUDITS

4.1. The parties hereby agree that CONDUCTWELL shall be permitted, at CONDUCTWELL's expense, to have a mutually agreed upon independent certified public accountant audit each royalty report submitted by INSULATE to CONDUCTWELL, within six (6) days from the date it is received by CONDUCTWELL. INSULATE shall make its records available to said accountant and cooperate by providing all available records essential to the verification of the report being audited and said accountant shall maintain confidential all information learned in the course of examining INSULATE's records, with the exception of a report to CONDUCTWELL with the accountant's findings as directly related to INSULATE' obligations to make royalty reports and payments. In the event of a finding by such accountant of a material variance with the report issued by INSULATE, then INSULATE shall reimburse to

CONDUCTWELL the audit costs paid to the accountant and pay to CONDUCTWELL any additional royalties determined to be due.

5. PATENTS

5.1. Should INSULATE obtain information that any patents owned by CONDUCTWELL are or may be infringed, it shall provide such information to CONDUCTWELL, but shall have no further responsibility or obligation. Any patents obtained by INSULATE relating to the Licensed Goods or improvements to the Licensed Goods shall be the property of CONDUCTWELL. INSULATE shall promptly review any papers and execute, acknowledge and deliver all such papers as may be necessary or desirable, in the sole discretion of CONDUCTWELL, to obtain or maintain patent protection for the Licensed Goods and to confirm the ownership of any such patent by CONDUCTWELL.

6. TRADEMARKS

6.1. INSULATE shall use the trademark "ALWAYS USE GOOD CONDUCT", in such form as specified in writing by CONDUCTWELL, on the Licensed Goods in the Licensed Territory. INSULATE shall also use in connection with the trademark a "TM" or, where U.S. Federal Trademark Registration has been obtained, an "®". Once approved, INSULATE shall not depart from the approved form of the "ALWAYS USE GOOD CONDUCT" mark on any materials requiring approval without the approval of CONDUCTWELL in accordance with paragraph 6.4 of this Agreement.

6.2. In order to assure the development, manufacture, appearance, quality and distribution of the Licensed Goods is consonant with the quality of the trademark, CONDUCTWELL retains the right to review the Licensed Goods.

6.3. INSULATE shall submit to CONDUCTWELL for approval samples of all Licensed Goods prior to any distribution or sale thereof by INSULATE.

6.4. Any such submission of the Licensed Goods for approval which is not disapproved within fifteen (15) days shall be deemed approved. Any disapproval by CONDUCTWELL shall be submitted to INSULATE in writing within the aforesaid fifteen (15) days together with remedial changes which would remedy such disapproval.

7. ENFORCEMENT

7.1. If CONDUCTWELL obtains patent protection for the Licensed Goods, and any such patent is infringed by a third party, INSULATE and CONDUCTWELL may take appropriate action to suppress such infringement. As patent owner, CONDUCTWELL shall have the first right, but not the obligation to take action. In the event that CONDUCTWELL takes action against an alleged infringer,

CONDUCTWELL shall be entitled to the entire recovery. If INSULATE requests CONDUCTWELL in writing to suppress any infringement, identifying in the request the infringer and the circumstances of the infringement, and CONDUCTWELL fails to file suit against the identified infringer or to otherwise take action to cause the identified infringement to cease within sixty (60) days of INSULATE' request, INSULATE shall have the right to file suit against and to negotiate and enter into a settlement with the identified infringer. If INSULATE files suit, CONDUCTWELL is under no obligation to bear any cost of the suit. CONDUCTWELL, at INSULATE' expense, shall join in the suit and render assistance and sign all papers, as may be reasonably required in connection with such enforcement. CONDUCTWELL shall be entitled to 20 % of any court awarded or lump sum recovery, less costs, as a result of any enforcement of such patents by INSULATE.

7.2. If INSULATE becomes aware of a third party infringement of the "ALWAYS USE GOOD CONDUCT" mark in connection with electronics products in the Licensed Territory, INSULATE shall provide notice and the details of such infringement to CONDUCTWELL, and CONDUCTWELL shall take appropriate action to suppress such infringement.

8. TECHNOLOGICAL SUPPORT, QUALITY CONTROL AND PERFORMANCE ASSURANCE TESTING

8.1. CONDUCTWELL shall provide instructions and know-how for the production of the Licensed Goods.

8.2. CONDUCTWELL and INSULATE agree to jointly develop a Quality Assurance Plan to meet all applicable agency regulations and certifications for the Licensed Goods. INSULATE shall test each production batch of the Licensed Goods and shall maintain test records in accordance with the Quality Assurance Plan.

9. PRODUCT LIABILITY AND WARRANTY CLAIMS

9.1. INSULATE shall assume all liability for all claims of any nature with respect to the Licensed Goods distributed or sold by INSULATE. INSULATE hereby agrees to indemnify, defend and hold CONDUCTWELL harmless, from and against any loss, liability, damages and expenses (including reasonable attorney's fees and expenses) which may be incurred or for which CONDUCTWELL may be obligated to pay or for which CONDUCTWELL may become liable or be compelled to pay in any action, claim or proceeding against INSULATE and/or CONDUCTWELL for or by reason of any acts, whether of omission or commission, that may be claimed to be or are actually committed or suffered by INSULATE in connection with INSULATE' performance of this agreement. The provisions of this paragraph and the obligations under the same shall survive the expiration of this agreement. INSULATE shall maintain and procure at INSULATE' expense a comprehensive general liability policy including, but not limited to, contractual advertising and products liability coverage's with a policy limit of not less than $_____ million per occurrence. Such policy shall be in full force during the entire term of this agreement and shall be placed with a responsible insurance carrier and shall name CONDUCTWELL as an additional insured and provide for at least thirty (30) days prior written notice to CONDUCTWELL of the cancellation or modification of such policy.

9.2. In the event that INSULATE does not obtain and maintain the aforesaid policy continuously in effect, upon prior written notification to INSULATE, CONDUCTWELL may obtain such insurance policy on behalf of INSULATE. All premiums for such insurance policy shall be deducted from royalties due under paragraph 2 of this agreement.

10. TERM AND TERMINATION

10.1. This agreement will have an initial term of seven (7) years from the execution date.

10.2. This agreement shall automatically be renewed for subsequent five (5) year terms, subject to the right of either party to terminate upon written notice one (1) year prior to the expiration of the initial term or any subsequent renewal term.

10.3. Notwithstanding the aforesaid, this agreement shall be subject to the rights of earlier termination by the party indicated as hereinafter set forth:

(a) By CONDUCTWELL in the event that INSULATE fails to make royalty payments following ten (10) days prior written notice and demand to cure from CONDUCTWELL; provided, however, if INSULATE cures such default within the ten (10) days period then such notice shall be of no force and effect; or

(b) By either party in the event that the other party breaches any other material obligation imposed upon it under this Agreement and fails to cure such breach within a period of thirty (30) days after notice and demand for cure from the party not in breach; provided, however, if the defaulting party cures its breach within the thirty (30) days period, then such notice shall be of no force and effect; or

(c) By either party immediately upon the other party becoming bankrupt, insolvent, making an assignment for the benefit of creditors, applying for or consenting to the appointment of a trustee or receiver or if bankruptcy proceedings are instituted against INSULATE.

11. RESOLUTION OF DISPUTES BETWEEN THE PARTIES

11.1. This agreement shall be deemed entered into the Commonwealth of Pennsylvania and shall be construed and governed solely by the laws of Pennsylvania.

11.2. In the event of any dispute, difference or question arising between the parties in connection with this Agreement or any clause or the construction thereof, then and in every such case, unless the parties concur to the appointment of a single arbitrator, the difference shall be referred to three (3) arbitrators; one to be appointed by each party, and the third being nominated by the two so selected by the parties, or if they cannot agree on a third, by the American Arbitration Association. In the event that either party within one month of any notification made to it of a demand for arbitration by the other party, shall not have appointed its arbitrator, such arbitrator shall be nominated by the American Arbitration Association. The arbitration shall take place in Philadelphia, Pennsylvania. The arbitrators must base their decision with respect to the difference before them on the contents of this Agreement and the attachments thereto, and the decision of any two of the three arbitrators shall be binding on both parties. The arbitrators shall apply the law of the Commonwealth of Pennsylvania.

12. PUBLIC STATEMENTS. Any public statements or publicity concerning the existence or contents of this Agreement shall be subject to review and approval by the other party. CONDUCTWELL and INSULATE will consult with each other concerning the means by which customers and potential customers shall be informed of this Agreement.

13. GOVERNMENTAL APPROVALS. INSULATE will at its own expense apply for and obtain the approvals of such governmental and regulatory entities as necessary to the marketing and sale of the Licensed Goods in the Territory. CONDUCTWELL will furnish necessary technical support to assist in obtaining any such approvals.

14. GENERAL PROVISIONS

14.1. This agreement sets forth the entire agreement and understanding between the parties hereto relating to the subject matter hereof, and supersedes any prior or contemporaneous oral or written representations, inducements or promises not contained herein.

14.2. No amendment or modification of this agreement shall be valid or binding unless the same shall be made in writing and signed on behalf of each party by a duly authorized representative.

14.3. This Agreement and all rights and obligations herein shall be binding upon and inure to the benefit of and be enforceable against the parties and their successors or assigns. CONDUCTWELL and INSULATE shall make no assignment, pledge or hypothecation of this agreement or its performance thereunder without the express written permission of

CONDUCTWELL and INSULATE.

14.4. The failure to enforce any of the terms and conditions of this agreement by either of the parties hereto shall not be deemed a waiver of any other right or privilege under this agreement or waiver of the right thereafter to claim damages for any deficiencies resulting from any misrepresentation, breach of warranty, non-fulfillment of any obligation of any other party hereto.

14.5. If any term or provision of this agreement is held to be invalid or unenforceable by reason of any rule of law or a public policy, this agreement shall be deemed amended to delete the term or provision so held to be invalid or unenforceable therefrom and all other remaining terms and provisions of this agreement shall remain in full force and effect. Provided, however, if the invalid or unenforceable provision contains a material term or condition of this Agreement then either party shall have the right to terminate upon 5 days prior written notice following the determination of such invalidity or unenforceability. If any provision is inapplicable to any circumstance, it shall nevertheless remain applicable to all other circumstances.

15. NOTICE

15.1. Any notice or statement by any party shall be deemed to be sufficiently given when sent by receipted facsimile with a copy by prepaid, trackable overnight delivery, to the notified party at its address set forth above and to its counsel. These addresses shall remain in effect unless another address is substituted by written notice.

[Add Names and Addresses]

In witness whereof, the parties hereto have caused this agreement to be signed, sealed and delivered on the date indicated above.

Insulate Technologies, Inc.

Date: _____

BY: _____
Name:
Title:

ConductWell, Inc.

_____

BY: _____
Name:
Title:

Index

G. B. Halt, Jr. et al., *Intellectual Property in Consumer Electronics, Software and Technology Startups*, DOI: 10.1007/978-1-4614-7912-3, © Springer Science+Business Media New York 2014

Printed in the United States
By Bookmasters